Faisal Durrani

OPTIMUM CONTROL OF NATURAL VENTILATION

Faisal Durrani

OPTIMUM CONTROL OF NATURAL VENTILATION

How to make Natural Ventilation more effective

VDM Verlag Dr. Müller

Impressum/Imprint (nur für Deutschland/ only for Germany)
Bibliografische Information der Deutschen Nationalbibliothek: Die Deutsche Nationalbibliothek verzeichnet diese Publikation in der Deutschen Nationalbibliografie; detaillierte bibliografische Daten sind im Internet über http://dnb.d-nb.de abrufbar.

Coverbild: www.ingimage.com

Verlag: VDM Verlag Dr. Müller GmbH & Co. KG
Dudweiler Landstr. 99, 66123 Saarbrücken, Deutschland
Telefon +49 681 9100-698, Telefax +49 681 9100-988
Email: info@vdm-verlag.de

Herstellung in Deutschland:
Schaltungsdienst Lange o.H.G., Berlin
Books on Demand GmbH, Norderstedt
Reha GmbH, Saarbrücken
Amazon Distribution GmbH, Leipzig
ISBN: 978-3-639-36442-2

Imprint (only for USA, GB)
Bibliographic information published by the Deutsche Nationalbibliothek: The Deutsche Nationalbibliothek lists this publication in the Deutsche Nationalbibliografie; detailed bibliographic data are available in the Internet at http://dnb.d-nb.de.

Cover image: www.ingimage.com

Publisher: VDM Verlag Dr. Müller GmbH & Co. KG
Dudweiler Landstr. 99, 66123 Saarbrücken, Germany
Phone +49 681 9100-698, Fax +49 681 9100-988
Email: info@vdm-publishing.com

Printed in the U.S.A.
Printed in the U.K. by (see last page)
ISBN: 978-3-639-36442-2

Acknowledgements

I would like to express my deepest gratitude to my research advisor Prof. J.A.Wright for his guidance and constant support in helping me to conduct and complete this work. I would also like to thank Prof. Malcolm Cook for his expertise on Natural Ventilation. I owe my sincere appreciation to my friends who have supported me and encouraged me over the years. I want to extend my profound appreciation to my beloved parents for their love, affection and invaluable support during my life and studies. I would like to also thank my brother and two sisters who have made my life as colorful as it could be. Finally, I would like to thank my wife for her love and understanding.

Abstract

Natural ventilation is one of the strategies used by the UK to reduce carbon footprint from buildings. Yet poor decision making during design stage can cause natural ventilation to fail in meeting its goals.

The overall goal of this research effort was to optimize the vent opening areas for a selected building. To achieve this goal, a detailed three dimensional virtual building model was created using DesignBuilder while EnergyPlus was used to study its thermal performance. Genetic Algorithm was used to optimize the vent opening areas to meet the overheating criteria while reducing the carbon footprint. The relationship between the vent opening areas and energy consumption was studied. A sensitivity analysis was carried out on the inlet and outlet vents separately. Other factors such as the population and evaluation size for Genetic Algorithm were also studied.

An optimized configuration of inlet and outlet vents was achieved with reduced carbon footprint and better thermal performance. The success of Genetic Algorithm as an optimum designing tool will enable the replacement of the commonly used rules of thumb in sizing vent openings. Further improvements can be achieved if Genetic Algorithm is used in designing the modulation strategy of vents.

Table of Contents

List of Figures:

List of Tables:

List of Acronyms, Abbreviations and Symbols:

ACH	Air changes per hour
BOP	Building Optimization Program
BRE	British Research Establishment
C.A	Cross-sectional area
CIBSE	Chartered Institute of Building Services Engineers
GA	Genetic Algorithm
IES	Integrated Environmental Solutions
^{o}C	Degrees Celsius
CO_2	Carbon Dioxide
kWh	Kilo Watt hours
%	Percentage sign

Chapter One - Introduction

1.1. Introduction

The emissions of greenhouse gasses have been one of the primary concerns of the UK. One significant area where UK has played a crucial role is in the area of reducing carbon footprints from buildings by implementing natural ventilation. Usually, catastrophic failure of naturally ventilated buildings is a result of poor decision making at the design stage. The way air flows in a naturally ventilated building is complex and depends significantly, amongst other factors, upon the shape and size of both the inlet and outlet openings. With the correct designing of these openings, summer time overheating and high heat loss during winters can be avoided.

1.2. Background and Motivation

Today, the construction industry is changing its design process. Architects and engineers work side by side and as a team from the early stages such as the conceptual design to the final stages known as detail design. The successful working of a mechanical ventilation system is not greatly dependent on the buildings they are installed in. On the other hand, natural ventilation systems need to be designed together with the building. The design of the building and its components, such as vents, are the elements that define the type and amount of air flow that will take place within the building. Apart from that, the design of vents will also influence other factors such as dust, noise and pollution ingress into the building.

I came across a briefing document for a student modeling competition organized by IBPSA. The competition asked the students to use computer simulation to model the control of a hybrid ventilation system for a building situated in Glasgow with stack assisted cross ventilation. The competition brief stated all the dimensions and properties of the building but the opening areas for natural ventilation were left up to the students to decide. The competition brief did however give a guideline to use an opening area in the range of 1-2.5% of the floor area that was to be served. The building was meant to have hybrid ventilation but the brief still asked the students to aim for providing natural ventilation for as much of the occupancy period as possible. The competition also stated that the dry resultant temperature of 28°C was not to be exceeded for more than 1% of the occupied period.

What interested me was the fact that the floor area was $561m^2$. This meant that students could size each opening area from about $1.5m^2$ to $4m^2$. This range of allowable variation would not only affect the area taken by vent openings on the walls of the building but would also significantly affect the type of air flow inside the building. Any student who would size these openings incorrectly could probably end up facing the following consequences:

- Natural ventilation limited to smaller duration.
- Higher energy consumption due to mechanical ventilation being turned on prematurely.
- Higher temperatures prevailing inside the building if vents are undersized.
- Higher heat loss during winters if vent are oversized.
- Higher risk of noise, dust and pollutants infiltration if vents openings are oversized (not part of the competition criteria).

This is not so different from the practical world. Getting the sizes of these openings is important for making natural ventilation work in real buildings. Once buildings are built it is often unfeasible to make changes in the design. It is therefore very important to make the right decisions during the design stage as buildings have a lifespan of 50 to 100 years.

With all the above points indicating the significance of getting the size of openings right I was motivated towards investigating an optimum size for vent openings.

1.3. Overview of Natural Ventilation

Humans spend most of their time inside buildings as they work, live and recreate in them. Due to this reason buildings consume high levels of energy. Since the energy crisis in 1973, buildings were made more air tight and thermally insolated in order to achieve lower levels of energy consumption required for cooling and heating buildings. Today, an integrated approach is used which focuses not only on lower levels of energy consumption but also on sustainable technologies such as natural ventilation.

Natural ventilation is a sustainable, energy efficient and clean technology if it is correctly designed and controlled within a building. In winters natural ventilation is required to provide sufficient air exchange in order to remove pollutants inside the building and provide fresh air requirements for occupants. While in summers an additional requirement is to maintain internal temperatures within comfort levels. Naturally ventilated buildings have a reputation of having fewer sick building syndrome (SBS) cases when compared to buildings having mechanical ventilation. Natural ventilation is also favored by occupants as they are usually in control of their thermal environment.

Presently, there are high-performance computers capable of intelligent building modeling, simulations and optimization and thus can help us in making better design decisions.

1.4. Research Goals and Approach

This thesis presents a parametric study in order to optimize the openings in a naturally ventilated building. The optimization is conducted on a building that was part of a student modeling competition and aims to meet the criteria set by the competition brief. The objectives of this research are:

- To know if natural ventilation is needed and can work without the assistance of mechanical ventilation
- To come up with an optimization algorithm which can optimize the vent openings in the building model
- To study the behavior of the optimization algorithm under different set of conditions
- To understand the relation between the cross-sectional area of the vent opening and energy consumption of the building
- To optimize the vent openings in order to reduce energy loss during winters
- To optimize the vent openings in order to avoid summer time over heating

This research has a deductive approach. By a deductive approach it means that we first decide upon a theory or hypothesis. Our hypothesis is that there must be an optimum size of vent opening between the recommended 1-3% opening area range, which would not only meet the temperature requirements during summer but also result in lower energy loss during winter. Based on our hypothesis a series of simulations were executed in order to get the confirmation if our hypothesis was correct or not. For this reason a building model was set up in DesignBuilder and an optimization tool known as the Genetic Algorithm was used to optimize the vent openings based on our criteria. The results later would validate out hypothesis.

1.5. Book Outline

This book is based on five chapters including the necessary background, motivation and the objectives of the research explained in Chapter 1. Chapter 2 gives some information on natural ventilation and looks at previous research done in the same area by other researchers. An overview followed by a detailed explanation of the methodology used to achieve the goals of this research is presented in Chapter 3.

Chapter 4 covers the thorough analysis and discussions of the results obtained from numerous simulations as a result of following the methodology. Questions are raised and their answers are investigated to fully understand the optimization process of the vent openings. And finally, Chapter 5 provides conclusions and a summary of research effort done for this project. Apart from this, some recommendations for future research interest are also stated.

Chapter Two - Literature Review

2.1. Introduction

UK has dedicated itself towards reducing energy consumption and carbon footprint from buildings. So the question arises how does a Building Services engineer help? The answer is simple. Throughout the life of buildings, the building services installed in them have to operate to make them comfortable to work and live in. The operating cost of such services is a big proportion of the whole life costs of the building. According to a research by Evans et al. (1998); design, build and operating costs are in the ratio of 1:5:200.

So with a lesser energy consuming and 'greener' ventilation strategy such as natural ventilation this running cost could be taken care of as well. But installing natural ventilation strategy is not the complete answer and guidance is needed in the design of such systems. For example, under or over sizing the openings of air vents can either cause higher operating costs, draughts, noise and even provide insufficient fresh air and thus affect occupant's performance. Thus this project will look towards an innovative optimum control of natural ventilation.

Prior to carrying out the project, it is essential to review the literature that is present out there on this subject. This is done to avoid the risk of duplication and also to get familiar with:

- the amount of knowledge that is currently present about natural ventilation
- the experiments carried out by other researchers
- the way previous researchers have carried out research
- the tools they have used
- the results obtained from previous works
- the areas that still need more research.

But before we embark on the review (of some) of the literature that is present on natural ventilation, we will first brush up on our knowledge of what natural ventilation is and what are the different natural ventilation strategies that are in use.

2.2. Introduction to Natural Ventilation

2.2.1. Background

The concept of natural ventilation is not new and has been used for thousands of years around the world. Yet with time and human development it was replaced by air conditioning. Air conditioning has been the answer to many building's indoor air quality needs until recently when natural ventilation was 'resurrected'. This was done to tackle the problems that were faced by the construction industry due to air conditioning; such as high energy consumption, CO_2 emissions and Sick Building Syndrome (CIBSE AM10, 2005).

2.2.2. Definition

'Natural ventilation may be defined as the ventilation that relies on moving air through a building under the natural forces of wind and buoyancy.'

(CIBSE Guide B2, section 2.5.1, 2001)

2.2.3. Explanation

CIBSE AM10 (2005) explains that as natural ventilation works on those driving forces, that are under the control of nature and not humans, it is inevitable that buildings will have wider levels of indoor environment and close control will not be possible. Yet it is now being foreseen that strict levels of control may not be necessary for occupants to feel thermally comfortable, if the occupants are in control of their thermal environment. For example, being able to open and close windows.

Generally, moving air is brought into the building either by windows or purpose built ventilation openings in the building façade. Natural ventilation does limit rooms from being too far from external walls or in other words natural ventilation works best in shallow plan buildings. This also helps natural light to penetrate deep into the building space. From an open window on the building façade to 6m deep into the building plan, natural ventilation is considered to be effective thus by shallow plan we mean a building with a width up to 15m (considering windows on both sides of the building).

Over the years approximate depths were estimated for the following configurations to which natural ventilation is considered to be effective:

- Single-sided single-opening = 2 times floor-to-ceiling height
- Single-sided double-opening = 2.5 times floor-to-ceiling height
- Cross-ventilation = 5 times floor-to-ceiling height

If buildings are of a deeper plan than the mentioned above then one can install the following to help assist natural ventilation (Levermore, 2000):

- Towers
- An atrium/atria
- Streets
- Funnels or chimneys
- Windcatchers

2.3. Natural Ventilation Strategies

Natural ventilation is a broad term that incorporates many different strategies in which it can be installed in a building. That is why knowing only what natural ventilation is, may not be enough to understand the literature present out there. These different strategies will be talked about time and again in different literatures thus it will be better to familiarize ourselves with them prior to getting them confused during our review. CIBSE AM10 (2005) explains them in the following manner:

2.3.1. Single-sided ventilation

As the name implies, single-sided ventilations points to such a configuration where the opening(s) is only on one side of the ventilated space.

a) Single-sided single-opening ventilation

If there is only one opening in the space being ventilated then the ventilating air will not penetrate deep into the room and the ventilation rate will be low as well. This sort of strategy is known as 'single-sided single-opening ventilation' depends only on wind as the driving force.

b) Single-sided double-opening ventilation

On the other hand there is a configuration that has a higher ventilation rate and deeper penetration into the ventilated enclosure. This requires openings at different heights compared to one another on the same façade, and is called 'single-sided double-opening ventilation'. Having two different openings lets stack effect enhance the ventilation rate. The further apart the two opening are from one another the higher the stack-induced flows. Yet caution should be taken when positioning the openings as low level openings do have the tendency to create draughts in winters.

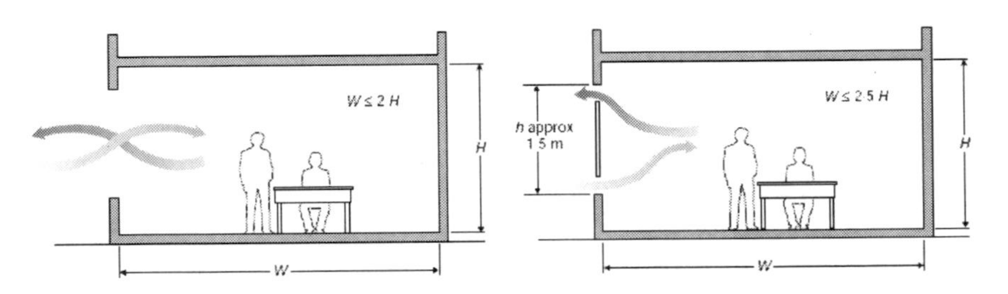

Figure 2.1: Single-sided single-opening ventilation (left), Single-sided double-opening ventilation (right). Source: CIBSE AM10 (2005) Pg: 15

2.3.2. Cross Ventilation

Cross ventilation is wind driven and takes place when ventilation openings are situated at two opposite sides of the enclosure. When wind drives ventilating air from one end of the enclosure to the other end, it flushes out any internal heat and pollutants present in the space. There are though two factors to keep in mind when using this strategy. Firstly, there is a limit to how deep the plan of the building can be to ensure cross ventilation to be effective. And secondly, any resistance to flow such as partitions must be low otherwise insufficient flow will occur. Cross ventilation does not have to use open windows or doors on either side alone, configurations such as a *'wind scoop'* or a *'duct/underfloor cross-ventilation'* can also be used where such applications would be more effective.

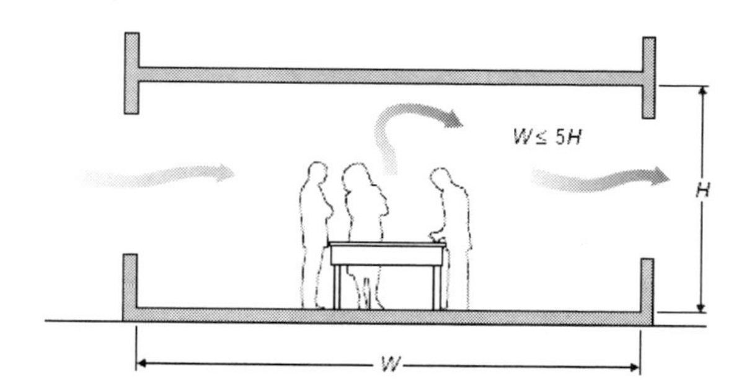

Figure 2.2: Cross ventilation, Source: CIBSE AM10 (2005) Pg: 16

2.3.3. Stack Ventilation

Stack ventilation exploits difference in the density of warm air (in a column) and the cold air surrounding it to make it work. As the warm air rises in the column, cooler air replaces it at the bottom. This action helps in producing a net outward flow of air from the building and fresh air replacing it. These columns or 'stacks' can induce high rates of flow if the outlet of the stack is in a region of negative pressure induced by wind. Also as air enters at the bottom of the building and is exhausted up, care needs to be taken when sizing the outlets of the stack to each floor in a multi storey building. Stale air from the building can be exhausted via a chimney or an atrium.

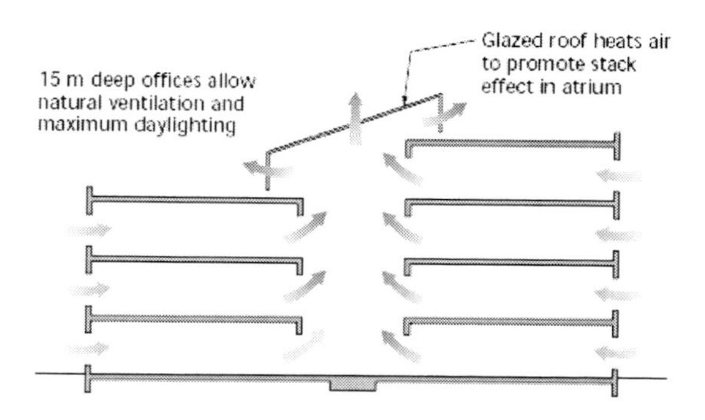

Figure 2.3: Stack ventilation, Source: CIBSE AM10 (2005) Pg: 18

a) Chimney ventilation

The role of chimneys is to maintain warmer air in them than the ambient and this becomes a problem as much of the chimney structure is exposed to the atmosphere. Thus chimneys need to be well insulated when being designed. This warm air is what helps in generating the stack effect. Their size depends upon the pressure drop requirements. A building may have one big chimney or a group of small chimneys depending upon the flow path required. To enhance the stack effect, chimney structures may have glazed elements at the top to allow solar radiation to heat up the air inside the chimney. These sort of chimneys are called 'solar chimneys' and result in increasing the overall buoyancy and hence ventilation rates. CIBSE AM 10 explains that it should be made sure that in winters the heat gain from solar radiation through the glazed elements is higher than the heat lost by conduction to the environment via the glazed fabric. Furthermore, proper designing of the roof profile of the chimneys can produce a negative pressure at the outlet which helps in making stale air flush out.

b) Atrium Ventilation

Atriums work similar to chimneys, yet an atrium serves many other purposes in a building as compared to a chimney. An atrium has the following additional benefits:

- It can be used as a social interactivity and occupancy circulation space in the building
- If placed in the center of the building it can draw air from both ends of the building to an extract at the center and hence making it possible to have twice the width of the plan as would have been used before.
- It can also be used to get in sunlight into the building hence reducing lighting demand.
- It can be used to reduce winter losses through conduction by acting as a buffer zone against the surroundings.

The roof of the atrium can also be fitted with fans to assist in the extraction of air on hot still days.

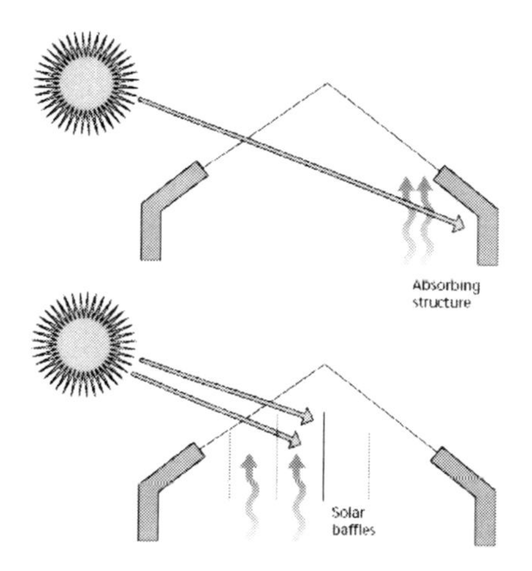

Figure 2.4: Atrium ventilation, Source: CIBSE AM10 (2005) Pg: 18

2.3.4. Double-skin façade ventilation

This type of ventilation is used where the building façades remains transparent to let in sunlight yet the resulting heat gain risk is taken care of by naturally ventilating the cavity and getting rid of the heated air. Thus we can say that the two façades act as solar chimneys. This type of strategy has the benefit of splitting tall buildings into smaller vertical stacks. Yet care needs to be taken when doing the thermal and lighting performance of the façade as well as condensation risk analysis and fire/smoke spread.

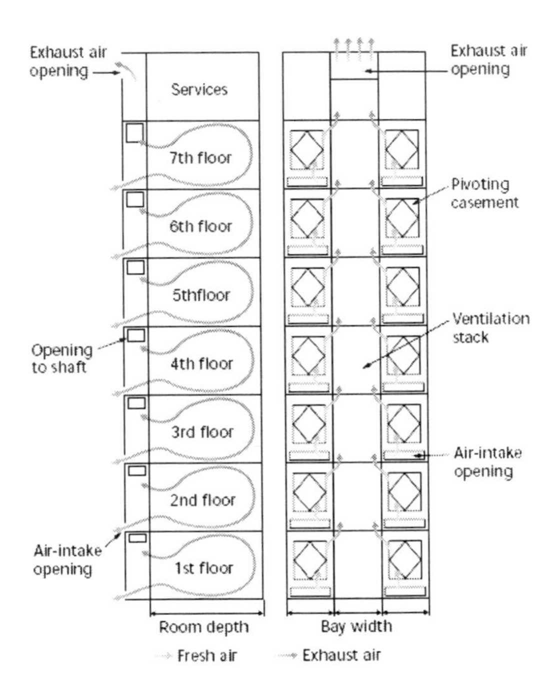

Figure 2.5: Double-skin façade ventilation, Source: CIBSE AM10 (2005) Pg: 19

2.3.5. Mechanically assisted strategies

The strategies mentioned previously, worked on 'natural means' most of the time and were supported by 'mechanical means' during hot extreme conditions. But mechanically assisted strategies are those which use mechanical means all the time during their time of run. The two types are:

a) ***Mechanical supply/natural extract:*** As the name implies, fresh air is supplied to the building by mechanical means and extracted naturally by chimneys or atriums. These are used where the building has a deep plan and the supply of fresh air in not possible to some sections of the building. Hence an under floor supply exists at different parts of the building and an extract at the top of the room. This type of strategy helps in reducing external noise and pollutants from coming into the building. The extract of air from the building at one single point suggests the air at the point of exhaust will be warm and hence can be used for heat recovery.

b) ***Mechanical Extract/natural supply:*** Mechanical extract/natural supply uses mechanical means at the exhaust while fresh air is sucked into the building due to lower pressures inside the building. This sort of strategy is beneficial where there is a high risk of internal pollutants such as printing rooms or smoking rooms. Mechanical extracts thus ensures getting rid of any internal pollutants.

2.3.6. Night Ventilation

During night time the temperatures drop and hence the inside-outside temperature difference is higher. This causes stack effect to increase and hence the flow rates increase as well. Apart from this the cool air at night cools down the building's fabric and hence on the following day the radiant temperature of the building is low. Again care is needed not to cause over cooling which can result in discomfort on the following day as well. Ventilating the building at night helps avoid risks of draughts and noise during occupancy timings. But naturally ventilating the building during night time does result in security risks for which purpose special opening limiting devices, dampers and louvers can be used.

2.4. Literature Review

2.4.1. Natural ventilation strategy study

To understand the limiting conditions of natural ventilation, Jackman (1999), carried out both physical and computer modeling for single-sided ventilation (summer time) on an office and came up with results. After the analysis of his results, Jackman was able to provide guidelines to single-sided natural ventilation design. Some of his guidelines were:

- Average indoor temperatures reduce with increasing window openings as a consequence of increased ventilation rates.
- Local ventilation rates and air temperature do not vary significantly with distance from the windows.

- The requirement of 8liters/person or 1.3liters/m^2 of the floor area (BSI, 1991) can easily be met if the window opening is in a ratio of about 1:20 with that of the floor area.
- However the goal to meet is thermal comfort and not ventilation rate and hence if internal gains within a building exceed 10W/m^2 then higher opening areas will be required.
- Single-sided ventilation can work efficiently in offices having a plan of about 10 to 12 meters.
- The flow of air in single-sided ventilation is such that cooler air enters from outside at low level, travels across the floor area until pulled up by heat generated from equipment and heat plumes from occupants. This warm air then travels at the ceiling height, finally exiting at the higher level of the window.
- Centrally pivoted windows provide better ventilation rates as compared to top-hung windows.

To validate the fact the stack effect is the dominant factor in causing ventilation in a single sided ventilation control strategy, Eftekhari et al. (2003) carried out experiments in which she measured airflows and temperatures at different locations in a test room. She then ran simulations with the help of the ROOM and VENT software to confirm that both the practically measured and software results were in agreement to one another. The results clearly showed a temperature gradient existing in the test room with the temperature at the lower levels being lower than the temperatures at higher levels. Her investigation also validated the fact that if internal temperatures are higher than the external temperatures then air will enter the space through the lower levels of openings and exit though higher levels of openings.

A special case of Single-sided ventilation was investigated by Jiang and Chen (2003). They used computational and experimental experiments, which were carried out in order to study in detail how single-sided ventilation is affected by buoyancy alone. For this sort of experiment it was necessary to neglect the effect of other variables such as wind. For this reason a day in summer without wind was considered as the experimental condition. Empirical methods were then used as a tool to measure the effectiveness of the ventilation in such an environment. The computational fluid dynamics (CFD) model used the following two methods in order to see if CFD simulation was suitable for such a situation:

- (LES): Large Eddy Simulation with filtered Dynamics Subgrid-scale Reynolds stresses (FDS).
- (RANS): Reynolds Avereged Navier-Strokes equation with experimental measurements of air temperature, velocity and ventilation rates.

The results showed that the RANS method had accurate and precise results when compared to the LES method. Furthermore, the RANS method also had better results than the empirical methods used prior to CFD modeling.

In context to the above the above mentioned study, Evola & Popov (2006) carried out wind driven natural ventilation experiments using CFD simulation. The experiments investigated with the help of the RANS method three situations, namely:

- Single sided ventilation with an opening on the windward side
- Single sided ventilation with an opening on the leeward side
- Cross ventilation through a room

The results were to be compared with the wind tunnel testing of a scale model of a room:

- With the dimensions: 2500mm x 250mm x 250mm (length x width x height)
- With an opening dimensioned as: 84mm x 125mm (width x height).

The opening was positioned on the windward side, leeward side and then on the opposite walls of the room (cross ventilation).

The RANS method itself consisted of two methods which were the Renormalization Group (RNG) theory and the two equation k-ε model. After getting the results it was concluded that having opening on the leeward side resulted in higher ventilation rates as compared to having them on the windward side. This though, the author advised, needed to be further investigated.

Sandberg (2004) carried out a study on the mechanism of cross ventilation by carrying out both wind tunnel testing and CFD simulation using Reynolds stress model and turbulence model. The study describes that in a building having an opening, the flow has a choice to flow through the opening or pass around the building. If the latter happens then, this results in a flow tube passing through the opening. In his paper he defines a new phenomenon known as the 'catchment effect' or the 'attractor effect' which in certain cases and conditions causes higher flow rates than the reference flow rate.

To summarize his findings, one can say that a problem of finding ventilation rates for cross ventilation is similar to knowing the pressure and velocity at each end of the openings but not knowing the tube section area. He explains that to get a beginning approximation one can assume that the sizing of the flow tube is being measured at the starting point and assume that the catchment area is equal to the opening area. Then if we know the reference velocity, finding the flow rates is possible. His work comprised of many opening sizes and it was seen that if the catchment area is 80% of the opening area the results are very much precise.

Cross ventilation in a building with non-symmetrical openings was investigated by Stavrakakis et al. (2008). The subject that was chosen to carry out the investigation on was a full scale model of a building in Greece for two hot summer days. For his investigation he chose to carry out a CFD simulation and compare it with the results he had obtained from experiments to see if they were valid. He used the following the RANS models for his CFD simulations:

- the k-ε model
- the RNG k-ε model
- and a 'realizable' k-ε model

The results (from all three models) when compared to the measured results showed good agreement with one another, thus validating the fact that the above three RANS models could be used for the analysis of cross ventilation in buildings with non-symmetrical location of openings. Another point to note was that a ±10°C change in wind speed or a ±15°C change in wind direction didn't affect the internal air flows significantly. A conclusion was made that having openings in the building in non-symmetrical locations provided good internal mixing of the air. As a bonus, having non-symmetrical locations of windows resulted in the reduction of draughts and high temperature gradients.

A design tool was created by Shaviv et al. (2001) to calculate the effectiveness of night ventilation. This was done with the help of carrying out simulations on the ENERGY software on an apartment building situated at four different climatic locations. Apart from that, they also had four different thermal mass levels and four different ventilation rates. Their results were quite as one would expect. As the ventilation rates at night time cooling were increased, the maximum indoor temperature on the following day would decrease. Yet this relationship is seen for a maximum of 20ach, above which one does not see any significant reduction in the indoor temperature. A conclusion was made based on the simulation results, which was that; higher the temperature swing in a region, the more indoor temperature can be reduced for the following day.

2.4.2. Natural ventilation design study

Windvents are devices that help to passively cool a building by creating a lower pressure in the room, which in turn draws in fresh air from the outside. Windvents regulate the ventilation rate with the help of dampers and hence it requires an investigation into what is the optimum opening angle for the dampers. This ensures lower running costs of the building as well as confirms the system to be in accordance with the legislation. Hughes & Ghani (2008) used FLUENT for his CFD modeling of the windvent for his investigation. He ran 19 simulations, where he kept boundary conditions constant such as keeping the external wind velocity equal to 4.5m/sec which is UK's average external wind speed. With such boundary conditions fixed, he ran simulations for different damper angles with a 5° increment.

After he ran the simulations he found two things:

- With the increase in damper angle, the pressure drop also increases. This is because as the damper angle increase the damper blades become more of an obstruction for air to flow and thus cause higher pressure drop.
- With the increase in damper angle, the velocity of air reduces. This is also because with the dampers slanting on higher angles, they restrict the flow of air and hence reduce the velocity of air flowing into the room.

However, the aim of the dampers control strategy is to minimize their movement, thus maintaining a constant delivery rate. By doing so for this experiment, an operating envelope around 48°, between 45° and 55° would increase the likelihood of optimum delivery conditions, and should be the target range for this type of application.

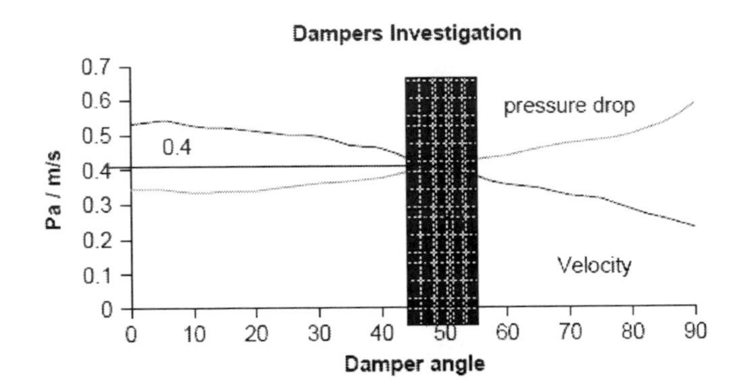

Figure 2.6: Effect on the velocity and pressure drop of windvent with increase in damper angle. Source: Hughes & Ghani (2008) Pg: 247

Heiselberg (2004) says in his paper that a good design is possible if there is integration between architectural design and the design of mechanical systems from the very beginning of the building's design. He explains the following two importance aspects of natural ventilation:

Natural ventilation design procedure

- ***Design of the building***

  Natural ventilation depends the outdoor climate and the thermal behavior of the building itself. The aim is to minimize heat gain in summers, maximizing heat gain and reduce the heat loss in winters and effectively make use of daylight and fresh air. Proper zoning of the building and night ventilation potential is also to be considered.

- ***Climatic design***

  Climatic design means taking benefits from the elements of nature for providing thermal comfort in a space. These elements can be sunlight, wind, earth, the air temperature etc. The key aim is to work with nature not against it. In this stage issues such as passive heating, passive cooling and daylight techniques are considered. Also design of and control strategies for natural ventilation are decided upon in this stage. Yet it is inevitable that there will be occasions when natural means may not be suitable for providing the required thermal comfort and hence comes the third step.

- ***Design of mechanical system***

  If there are any remaining loads that need to be catered for then mechanical systems are designed for them in this stage. E.g. exhaust fans or mechanical cooling systems.

Natural ventilation design challenges

Ventilation has usually two goals to achieve, control of air quality and natural cooling in summers. The ventilation rates are set to meet the indoor air quality requirements and keeping the energy use (during periods of heating and mechanical cooling) to a minimum.

Ventilation for control of air quality is done by:

- Reducing/isolating pollutants to reduce ventilation rate required, use of demand control and meeting fresh air requirement.
- Using heat recovery, passive heating and passive cooling to reduce loads.
- Maximize stack/wind effect and reduce losses in duct work to minimize fan load.

Ventilation for natural cooling requires optimization between cooling load and thermal comfort.

This is done by:

- Reducing external and internal heat gains such as equipment and lighting. And making use of sunlight and careful use of shading.
- Using thermal mass as a heat buffer which stores heat during occupied periods and releases it during unoccupied periods with the help of night ventilation.
- Maximize stack/wind effect and reduce losses in duct work to minimize fan load.

Fracastoro et al. (2002) did numerical modeling of the characteristics of air flowing through window openings (in a transient manner) such as air change rates, air temperature and ventilation efficiency. His analysis involved carrying out theoretical study by both CFD and zonal modeling and then comparing the results obtained, as well as comparing them with experimental measurements read from full scale models. The investigation included:

- CFD MODELS: consisting of a 2D geometrical transient model with a 2m wide strip of external environment. For the CFD model the following assumptions were made:
  - It's a standard k-ε turbulence model
  - Power-law interpolation scheme
  - Standard log-law wall functions
  - Number of iterations per time-step: 1000
- ZONAL MODELS: for the study of the zonal models the space was divided into:
  - Single-zone model
  - Two-zone model
  - Three-zone model
- NON-DIMENSIONAL ANALYSIS: with the assumption that $A=wH$, a non-dimensional equation was derived that expressed the flow rate of air through a single opening. This equation non-dimensionalized the independent variable time and the dependent variables air flow and temperature. The equation obtained was:

$$\dot{m} = w\mu C_d \sqrt{\frac{\rho^2 g H^3 (T - T_o)}{\mu^2 T}}$$

- EXPERMENTAL ANALYSIS: an experimental analysis was carried out in a real office with a single side opening. Temperature measuring sensors were installed as a grid throughout the space of the room. The results show that a warm air stagnant in nature forms over the upper edge of the window.

So in conclusion the unsteady behavior of air entering a room with an open window was studied in detail. For this reason different simulation methods were used in which zonal models were in agreement with the experimental results and the CFD readings were validated by comparing them with the zonal model results.

2.4.3. Natural ventilation control strategy study

Natural ventilation needs to be controlled for two purposes, explains Martin & Fitzsimmons (2000). Firstly, natural ventilation depends upon natural conditions and hence can vary drastically and thus needs to be controlled. And secondly natural ventilation is controlled in order to deliver the required ventilation rates with an occupied space. There are two ways in which natural ventilation is controlled:

*1. **Manual Control***

Naturally ventilated buildings are mostly controlled by having a local occupant control. Local occupant control can be occupants simply, opening windows manually to using push buttons to open and close windows. Yet there can be vents that are automatically operated (with an override capability) to cater for night cooling. Recent research has shown that occupants having a control of their environment are more tolerant to their surrounding temperatures than being in an automatically controlled environment (Oseland, 1994). Such that in an air conditioned building occupants may tolerate a maximum temperature of about 25°C while in a naturally ventilated building with the control in the user's hands, the comfort temperature can be up to 27°C.

*2. **Automatic Control***

Automatic control of the inlet and outlet of vents can allow for night time cooling to be carried out when the building is vacant during night time. Automatic control can also be integrated with sensors to control the ventilation rates to a building depending upon internal temperatures. If the building is being heated then ventilation rates can be kept low such that they just meet the CO_2 exhaust requirements. For automatic control there are components such as damper actuators, actuators and sensors in the building that help in the regulation of ventilation rates. Usually, automatic control of natural ventilation is part of the Building Energy Management System (BEMS); opening and closing many vents throughout the building at the same time.

There are certain issues of concern when it comes to the effectiveness of natural ventilation when applied to a building. A few of these issues are:

Restricted Ventilation and stagnant areas

Buildings might be designed in the best possible way for natural ventilation to take place in, but usually the desired ventilation rates are not achieved during the operation of the building. This is a result of either blockage of vents or internal arrangements of partitions and furniture. This causes difficulty in the ingress of fresh air into the building and also restricts it from reaching all the places in the building (requiring that fresh air).

Overcooling/Draughts

During winter season, when occupants open windows to let in fresh air and get rid of internal gains draughts and overcooling occurs. These draughts and overcooling of buildings are also difficult to control. It is therefore recommended that air velocities of 0.25m/sec in summers and 0.15m/sec in winters should not be exceeded in moderate thermal environments (BSI, 1995).

Non-Operation of vents

The purpose of ventilation openings is to provide ventilation rates that are sufficient to remove heat gains in summers and meet the indoor air quality requirements during winters. But the operation of these ventilation openings/vents is restricted when they get blocked or are closed completely due to some reasons. The reasons could be:

- Ingress of external noise
- Risk of wind or rain
- Poor external air quality
- Operational problems
- Human factors such as no-one 'owning' the window such as in public areas

Overheating

Increase in internal dry resultant temperatures causes overheating. Overheating is thought to happen only because if high internal heat gains. But factors such as high external heat gains, poor ventilation and general operational issues also cause overheating in buildings. Measures should be taken to reduce the major internal heat gains such as people, lighting, electrical equipment and solar gains.

Another interesting research was the comparison of results obtained from thermal resistance ventilation model (TRV) and that of the ESP-r program. Yao et al. (2005) stresses that ventilation strategy analysis should consist of two stages:

- The first step is to provide interactive design feedback, where the sizes of elements may be defined, different inputs may be validated and their results can be compared. Yet the accuracy of this stage is not high.
- The second step consists of finding the relative accuracies and finding how much does the change in one variable, deviate the end result from the original result.

The paper used a very simple control strategy for the automatic control of the natural ventilation openings. The control code was *(Where, T_I = internal room temperature, T_a = ambient temperature)*:

a. Wind = high > openings = closing
b. T_I = low, T_a = low > openings = closing
c. T_I = low, T_a = not low > openings = opening
d. T_I = high, T_a = high > openings = closing
e. T_I = high, T_a = not high > openings = opening
f. T_I = optimum, T_a = low > openings = closing
g. T_I = optimum, T_a = high > openings = closing
h. T_I = high, T_a = usual > openings = opening
i. T_I = high, T_a = low > openings = opening

The paper explains a method of design and assessment of natural ventilation which consists of the following two stages:

- Given air change rate profile (CAR)
- Dynamic air change rate (DAR)

These two stages can help in performing a more realistic and accurate thermal simulation and comfort assessment. The two stage analysis can also help in evaluating the effectiveness of the control strategy. A single zone method can be used to carry out examination of natural ventilation at the perimeter of buildings which usually have high solar and conduction gains from the external environment. For this case the heat transfer between the different zones within the building are neglected as they are small. Yet in cases, such as having cross ventilation carrying air from the south face of a building to the north face has high interactivity between the different zones of the building and hence the heat transfer amongst the zones cannot be neglected. If this is the case, then a multi-zone method can be used, which would result in a much accurate set of results. The main agenda of this paper was to talk about the proposed method and the process of its application.

Once it is decided to have automatic control for vents; that are out of reach or used for night time ventilation, a control strategy needs to be decided. These can be categorized as 'generic control strategies' and were defined by Martin (1996).

Natural Ventilation Temperature and CO_2 Control Strategy

This control strategy, as implied by the name, works on the temperature and CO_2 levels inside the ventilated space. The levels are measured by sensors and then compared with their respective setpoints. The difference between the setpoints and the values measured will control the opening levels of vents. The control signal controlling vent opening levels will be affected by the amount of wind speed, its direction, the external temperature and rain. This modified control signal will thus produce reduction in the opening levels of the vents and hence maintain comfortable conditions inside the ventilated space. This is will occur until higher levels of temperatures or CO_2 inside the building are reached.

Pre-Cooling Control Strategy No.1

This strategy maintains temperature equilibrium between the building fabric and the space by calculating the heat gains during daytime and getting rid of it at night time. The daytime heat gains in degree hours is defined as the hours the internal temperature was above the setpoint and similarly, cooling gains in degree hours are defined the number of hours the internal temperature is below the room temperature setpoint.

At the time when the building is being unoccupied, if the internal temperature for three degree hours is higher than the setpoint, also the outside temperature is lower than the room temperature then the building will be pre-cooled that night. Once this happens, vents then open and close to maintain a pre-cooling setpoint temperature inside the building and thus the cooling gains are measured. Pre-cooling session ends when night cooling gains degree hours are equal to daytime heat gains.

If sufficient night cooling does not occur then two things can be done:

- Turning off the pre-cooling earlier so that heat re-emitted from the building fabric can bring the space temperature to suitable levels.
- Pre-cooling can be prolonged and then the building can be heated up for a little time. It is believed that this provides additional cooling for later part of the day.

Pre-Cooling Control Strategy No.2

This strategy unlike strategy 1 utilizes the thermal capacity of the fabric, furniture etc to work. Strategy 2 measures the mean outside temperature from 12:00 to 17:00, and if for these hours the outside mean temperature is higher than the pre-cooling setpoint temperature and also the inside temperature is higher than the outside temperature then pre-cooling takes place. This is usually at the time the building is being unoccupied. Suppose the pre-cooling setpoint temperature is 18°C. As the vents are opened, cooling takes place until the inside temperature of 14°C is reached. Once this happens all the vents close and let the heat from the fabric of the building heat the space again to say about 17°C. Once again vents will open up to cool the building down. This cycle of cooling and heating occurs until the preheat period is reached. This is the point at which all vents have to be shut in order for the heat from the fabric to heat the building to the inside setpoint temperature, say 19°C.

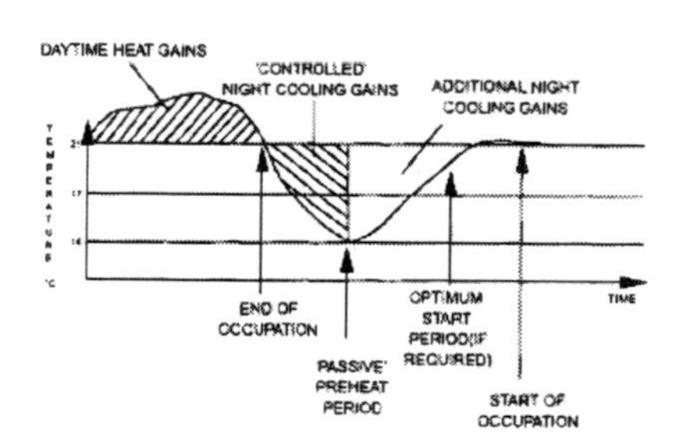

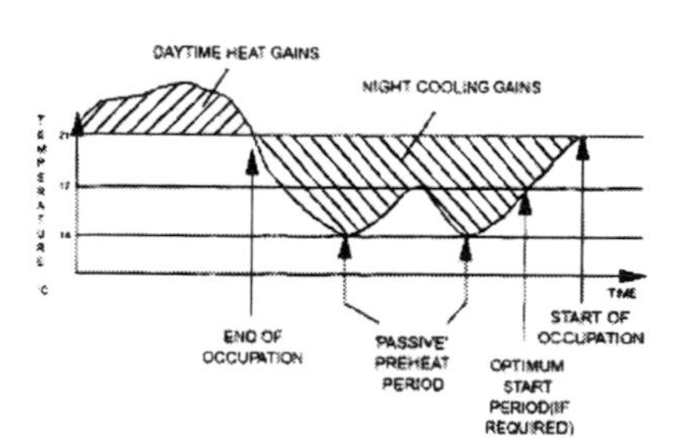

Figure 2.7: Pre-Cooling Strategy: No.1 (left), No.2 (right). Source: Martin (1996) Pg: 7

Mixed Mode Ventilation Control Strategy

For the execution of this strategy the heating setpoint and cooling setpoint temperatures are used. The average zone temperature is compared with the room heating setpoint temperature. If the average zone temperature is:

- 0.5 or more, higher than the heating setpoint > cooling mode.
- Within ±0.5 of the heating setpoint and pre-cooling was executed the previous night > cooling mode.
- Within ±0.5 of the heating setpoint and pre-cooling was not executed the previous night > heating mode.
- More than 0.5 lower than the heating setpoint > heating mode. The fans of the air handling unit turn on with the heating coil at the heating temperature setpoint. This is maintained until the room temperature is 0.5°C higher than the room heating setpoint temperature. At this point in time the fans turn off. Trickle ventilators and manually opening windows provide the rates of ventilation required in the space from this point onwards.
- Rises to more than 0.5°C above the cooling temperature setpoint, natural ventilation takes place.
- 1°C above the cooling setpoint then supply and extract fans of the air handling unit are turned on at lower speed but if the difference in temperature was up to 2°C then these fans would be running at high speeds. If still internal temperatures soar and reach a
- temperature of 26°C then mechanical cooling should be turned on with all the vents closed. Some vents may be kept open to maintain the minimum ventilation rates.

If there is an occurrence of rain, high winds or lower outside temperatures then all natural ventilation opening with are shut or kept to minimum levels.

A modified version of EnergyPlus and CFD was used by Da Graca et al. (2004) to come up with a control strategy for a naturally ventilated building. The building itself had cross ventilation acting through openings on each side of the building. Apart from that, on each side of the building there were two types of openings; the openings on higher levels were controlled by BEMS while the lower levels had a high number of user-controlled windows. The paper was based on an attempt towards having low energy indoor climate system or BEMS providing comfortable conditions inside the building. The pressure data measured on either end was what defined the leeward and windward side of the building in the control strategy. To model the behavior of the user in the experiment the following two types of users were defined:

- Uninformed users: this type of user behaves in a way that is not dependent on the BEMS actions.
- Informed users: this type of user is dependent on the BEMS actions. Also, when the BEMS is in the night cooling mode, it is designed that the users will also leave their windows open during that night.

It was seen that a main jet was formed at the upper openings (while spreading across the ceiling slab) from windward to the leeward side. While in the space below this jet (i.e. occupied space) was served with recirculating air. The results show that if the building is served with night time cooling in hotter seasons, with a suitable control strategy the building can have satisfactory comfort levels. The results also pointed to the fact that user behavior has high effect on the performance of the building.

Spindler & Norford (2008) explained that there is a huge effort being done in order to reduce the energy use by buildings, for which natural ventilation is one of the answers. Yet the control strategies that are in practice today only use heuristic strategies that are only dependent on indoor and outdoor conditions. These types of strategies, according to him, are 'sub-optimal' and lack the accountability of the dynamic thermal behavior of the building nor do they predict any future weather conditions.

The paper explains that zonal temperature can be predicted (in regards to different variables) with the use of both linear and non-linear data-driven models of buildings. These models can thus be used to reduce energy consumption while still meeting the recommended levels of thermal comfort in the building for the occupants.

For the test building (and a 24-hour period), the temperature control strategy employed was:

- To maintain the temperature within the upper and lower setpoint band in the assembly area of the building

- Use minimum electricity (in case of fan assistance)

The algorithm proposed in this paper proved its effectiveness for the control strategy optimization for a single day. The sensitivities of the optimized results (with fixed constraints) were analyzed by the Pareto front. The single-day optimization was based on concepts that also helped in generating algorithms for a multi-day case. Evidence was also shown to support the fact that these concepts could also be used in cases where that outside temperatures allow ventilation, and hence a supervisory control or a local control can be established. The paper also talks about how a linear model is not suitable for cases where the wind influence on the building is high. It was advised that another type of model is required in which the direction of the wind and other control modes may be addressed more clearly.

2.4.4. Other Work on optimum control

Spindler (2004) based a review of all natural ventilation control strategies on 'IF-THEN' statements, meaning the review of night cooling for naturally ventilated building was based on heuristics. But a point to note was that these heuristics were not optimized unless manually adjusted for a specific building. The review consisted of both supervisory and local controls for the natural ventilation optimization of the building.

Mahdavi & Proglhof (2008) used a test room with different types of openings for a wind driven ventilation case for which they found the air change rate with the help of a data-driven model. The air change rates were also found for a nodal numerical model and improved the predicted rates with the help of an empirical relationship with measured air change rates. Though the predictions were only a 'one hour ahead' simulations and didn't extend to longer periods, as would be necessary for night time cooling.

When it comes to night cooling then Heuristic automatic control is a better option than occupant control. This is because occupants can only set the louvers in specific positions and leave them unattended which can result in the ingress of cool or warm air into the building in excess. Leim et al. (1997), Passen et al. (1998), BRE (1998, 1999), CIBSE (2000) & Levermore (2000) provided engineers with detailed night time cooling decision trees. For the Inland

Revenue building in Nottingham, UK the night time cooling strategy employed was that vents shall open when:

- average outside temperature > 18°C (during noon and 5pm)
- outside temperature > 12°C
- inside temperature > outside temperature
- inside temperature > 15.5±1.5°C

Wright et al. (2002) with the help of the predicted percentage of dissatisfied occupants (PPD) came up with Pareto-optimal solutions that would pave ways to saving energy and reducing thermal discomfort. The Pareto-optimal solutions were based on use of a multi-objective genetic algorithm. Genetic algorithms were also used by Fanger PO (1972) to optimize the setpoint of the supply air for a building that is occupied from 8am to 5pm.

2.5. Airflow Network Model of EnergyPlus

An air distribution system's performance can be simulated with the help of an Airflow network model. Airflow network models make it possible to predict the airflows into different zones of a building due to effects of external winds and forced air by mechanical means (EnergyPlus, 2008). EnergyPlus uses this same airflow network model to predict airflows in different zones based on AIRNET (Walton 1989). The Airflow network model option of EnergyPlus incorporates the advantages of both COMIS and (Air Distribution System) ADS. The Airflow network model consists of linkages that connect nodes with the help of airflow components.

The Airflow network model follows the following three steps in sequence:

1. Calculation of nodal pressure and airflow: requires information about the wind pressures and air flows.
2. Calculation of nodal temperature and humidity: requires information about the zone air temperature and humidity ratios of the zones of the building.
3. Sensible and latent load calculations: uses the data of both zonal temperatures and humidity ratios to find out the sensible and latent loads for each zone of the building.

From the EnergyPlus software documentation discussed above, we understand that by defining nodes, linkages and airflow components in our own model we will be able to get the airflows into different zones of our building.

2.6. Conclusions

After the review the literature on natural ventilation it was seen that there exists a lot of literature on the subject, so much so that discussing it all in this chapter was not possible due to limitations on the literature that can be written. Natural ventilation is seen to have been researched a lot lately as it is seen as one of solutions to reducing energy consumption and carbon foot print from buildings.

We see that from the beginning of the 1990's, research has been done on understanding what natural ventilation really is and how, when and where is it applicable i.e. guidance to a good design of natural ventilation was discussed. These researches were mostly based on experimental, analytical and computational investigations. Once a good understanding was acquired about the design and control of natural ventilation, more and more buildings were installed with it. But very soon there was seen a need to improve how natural ventilation ran in buildings. Thus, from around 2005 to the present date numerous works are being done to optimize the control of natural ventilation which could further reduce energy use without sacrificing thermal comfort in buildings.

The optimization work done till now is mostly based on either:

- running CFD simulations to understand the flow of air (even in more detail) through the building
- optimizing algorithms (both heuristic and non-heuristic in nature) for a better schedule for vent opening.

Yet one very important aspect that is being overlooked or not being given its due importance is optimizing the vent openings and flow rates for different degree days of the year for an office building. Having improper vent opening sizes can result in higher energy use or poorer indoor air quality. Thus it would be helpful to investigate an optimum vent opening for a naturally ventilated building.

Chapter Three - Methodology

3.1. Introduction

The Oxford dictionary defines methodology as "a system of methods used in a particular field" or "a particular procedure or set of procedures". In scholarly literature methodology refers to the rationale and philosophical assumptions that support the scientific methods used to carry out the study. This chapter thus presents the research methodology selected for carrying out this research. It will be observed that the methodology was influenced by the purpose and objectives of this study. In the following sections of this chapter the purpose of the study will be reviewed, research questions and hypothesis will be presented and result analysis procedures and limitations of the project will be discussed.

3.2. Purpose of study

The purpose of this study is to examine the phenomenon of optimization of vent openings with the implementation of an optimization tool on a naturally ventilated building as a case study. The reason for doing so is to determine the optimum sizes of ventilation vents which avoid summer time overheating and result in smaller carbon foot print for naturally ventilated buildings.

3.2.1. Research Question

What are the optimum sizes of vent openings for the selected naturally ventilated building that provides temperatures below 28°C for as much of the time as possible without having higher energy consumption?

3.2.2. Hypothesis

The key factor to meeting the criteria of the competition is the correct design of vents that feed the building with fresh air. Our hypothesis is that once we run GA on our building model, GA will try to minimize the vent opening area in order to reduce energy consumption. With an additional constraint, GA will stop reducing the size of vent openings once there is a risk of the internal temperatures getting higher than 28°C for more than 31 hours. This is the vent opening which we will consider as an optimum size.

3.2.3. Research Problem

This research study, as clear from the research question, hypothesis and Chapter one examines a single problem. The problem is the correct sizing of vent openings for a naturally ventilated building. With the reduction of vent opening areas, the energy loss during winters is also reduced yet this causes higher internal temperatures during summers. On the other hand if the vent opening areas are increased not only do they result in higher energy losses but also results in problems such as noise ingress and pollution infiltration. Over sizing of vents mean the vent openings taking up more space on the walls inside the building. Another significant problem of over sizing vent openings is that to maintain comfort levels and avoiding draughts, louvers or dampers are used to regulate flow. Over sized vents cause higher air flows and so these louvers are kept at a slanted position in order to reduce the air flow. But if louvers are kept at slanted position for most of the time their lifespan is reduced. Having louvers and dampers in a slanted position also causes noise as air passes through them. Thus as explained, the research problem is finding the optimum size of vent openings that maintain temperatures lower than 28°C without over sizing the vent openings.

3.3. The Research Approach

Creswell (2003) suggested that the research approach influences design and an opportunity is given to the researcher to understand how various approaches may contribute or limit his research. The question here is to have a deductive or an inductive approach?

Marcoulides (1998) explained the two approaches. According to him a deductive approach tests a certain theory or a set of theories on the basis of which a hypothesis is formed. This hypothesis is then confirmed with the help of tests and observations. On the other hand, inductive approach starts to look at empirical findings or data and draws up concepts and theories.

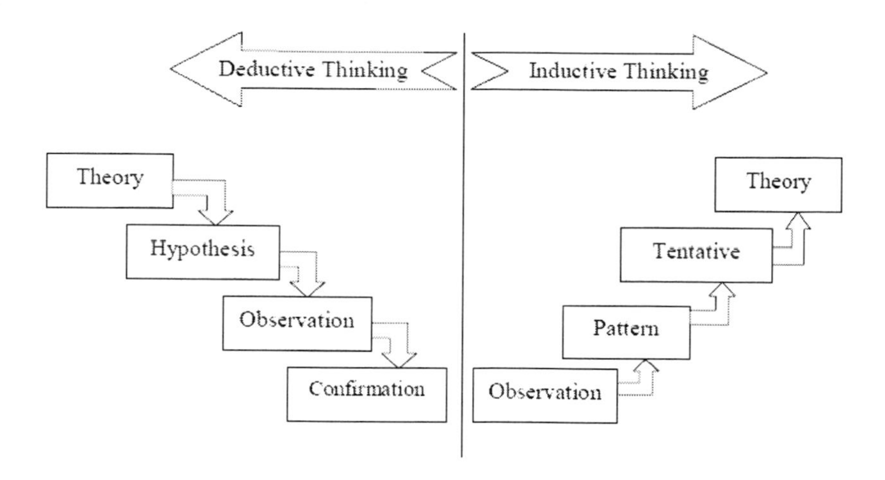

Figure 3. 1:Deductive Versus Inductive Approach. Source: Trochim (2001).

Figure 3.1 shows that deductive approach follows a top-down while the inductive approach follows the bottom-up approach. For this study the deductive approach appears more appropriate as it will test our theory and hypothesis discussed earlier. Let us now test our hypothesis with the help of some observations.

There are a number of procedures in which the optimization of vents could have been executed but the following procedure was chosen with the aim of covering all the objectives of this research. It is important at this stage to look at some of the information relevant to the execution of this research methodology before explaining the procedure followed.

3.4. Relevant Information

3.4.1. Building Description

The building used for the purpose of this study is taken from the student modeling competition briefing document which is a 3 storey building situated in the city of Glasgow (Scotland). The building consists of open plan offices, ventilated with the help of stack assisted cross ventilation. The plenum at the bottom is intended to provide fresh air to the central lightwell which will in turn deliver the fresh air to each level of the building. The fresh air will then flush out the stale air via four stacks provided on each face of the building façade. The dimensions of the building are shown in the section and floor layout of the building provided in figure 3.2. Other information about the building such as the activity, construction materials and openings are also given in the briefing document.

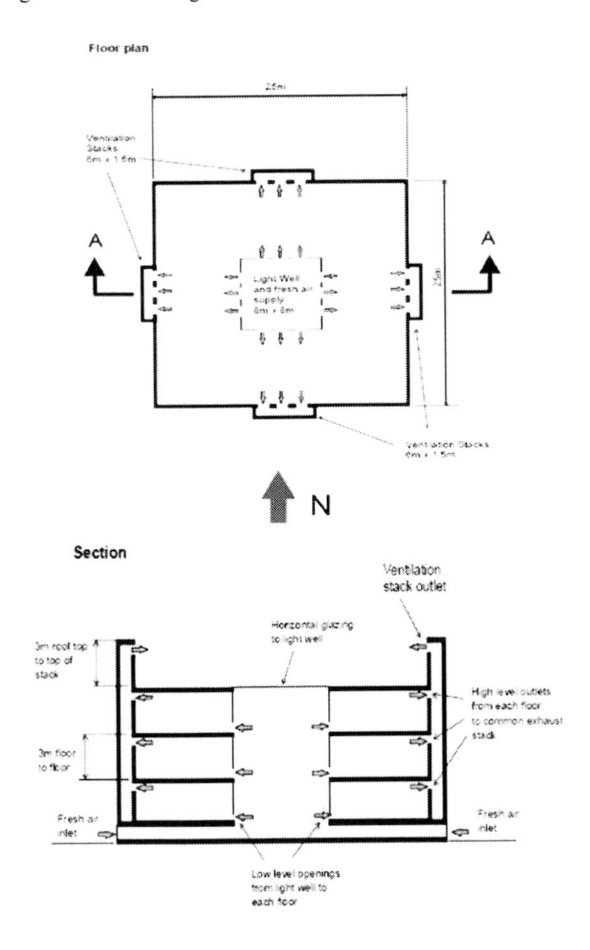

Figure 3. 2: Floor Plan and Sectional view of the Building

3.4.2. DesignBuilder

The Simulation engine used in this project was EnergyPlus which will be discussed in the next section. But it needs to be explained at this point that EnergyPlus is not a user-friendly package when it comes to modeling virtual buildings and usually Energy Plus is wrapped around by a third party interface. DesignBuilder software was chosen as this third party interface as it is user-friendly and can be used not only to construct virtual building models but also provide a range of environmental performance data. This environmental performance data is though, generated with the help of EnergyPlus.

DesignBuilder helps even the novice users to build virtual models with its easy to use OpenGL solid modeller in 3-D space. Building models can be assembled just by positioning, stretching and cutting 'blocks' which results in a hierarchy of blocks and zones of the building. Later on model parameters such as the building activity, construction, openings, lighting and HVAC can be added into the hierarchy with the help of separate tabs which the software provides to its users. These tabs mostly consist of lists, check boxes and scroll bars to assign the desired values. The modelling hierarchy is Site > Building > Block > Zone > Surface > Opening.

3.4.3. EnergyPlus:

There are many simulation engines available in the market such as Integrated Environmental Solutions (IES), EnergyPlus etc but commercial software such IES do not give access nor share their source code with their users. For this project a simulation engine was needed which could not only run simulations but also give us access into the source code. We needed this access so that we could introduce our optimization code into the source code later on during this project.

Energy Plus is an energy analysis and thermal load simulation program with the following two key benefits which made it highly suitable for this project:

- Open source code: EnergyPlus allows its source code to be available to its users for inspection and research purposes. This access to the source code was provided by its developers in order to improve the accuracy and usability of the program with time.

- Modularity: EnergyPlus is modular in nature and thus helps us to develop modules and introduce them into the code without interfering with other modules. Thus researchers are saved the efforts of learning about the entire structure of the code before being able to introduce their bit of the code.

3.4.4. EnergyPlus Launch:

The EnergyPlus Launch is the user interface that comes with the EnergyPlus software. The EP Launch requires, from the user, an input data file (IDF) and a corresponding EnergyPlus weather file (EPW) before it can run simulations. The EnergyPlus Launch also provides text editor for editing both input and output files, opens spreadsheets for viewing result files, opens a web browser for viewing tabular result files and can even open a viewer for drawing files. We will be using the text editor to edit the IDF files and introduce our optimization code into it.

3.4.5. Genetic Algorithm (GA) Optimization:

In mathematics optimization refers to choosing the best element from a set of available alternatives. Genetic Algorithm is a search technique used in computing to find exact or approximate solutions to optimization problems. In GA abstract representatives (chromosomes) of candidate solutions (individuals) to an optimization problem evolve towards better solutions. This evolution starts from a population of randomly generated individuals and takes place in generations. During each generation the following takes place:

- Each individual in a population is evaluated for its fitness
- Multiple individuals are stochastically (randomly) selected based on their fitness from the current population
- Multiple individuals are recombined or randomly mutated to form a new population.

This new population is then used in the next iteration of the algorithm. Usually the algorithm terminates when either a satisfactory fitness level is reached or when the maximum number of generations has been produced. The latter though may or may not result in a satisfactory fitness level. For this project a Genetic Algorithm code, which has been written by Professor J.A. Wright will be used.

3.5. Overview of the procedure

The methodology designed for the optimization of vent openings is presented by the following flowchart. It gives an overview of the series of steps taken in order to achieve the goal of this research. Following the flowchart is a detailed explanation of how and why each of steps was executed.

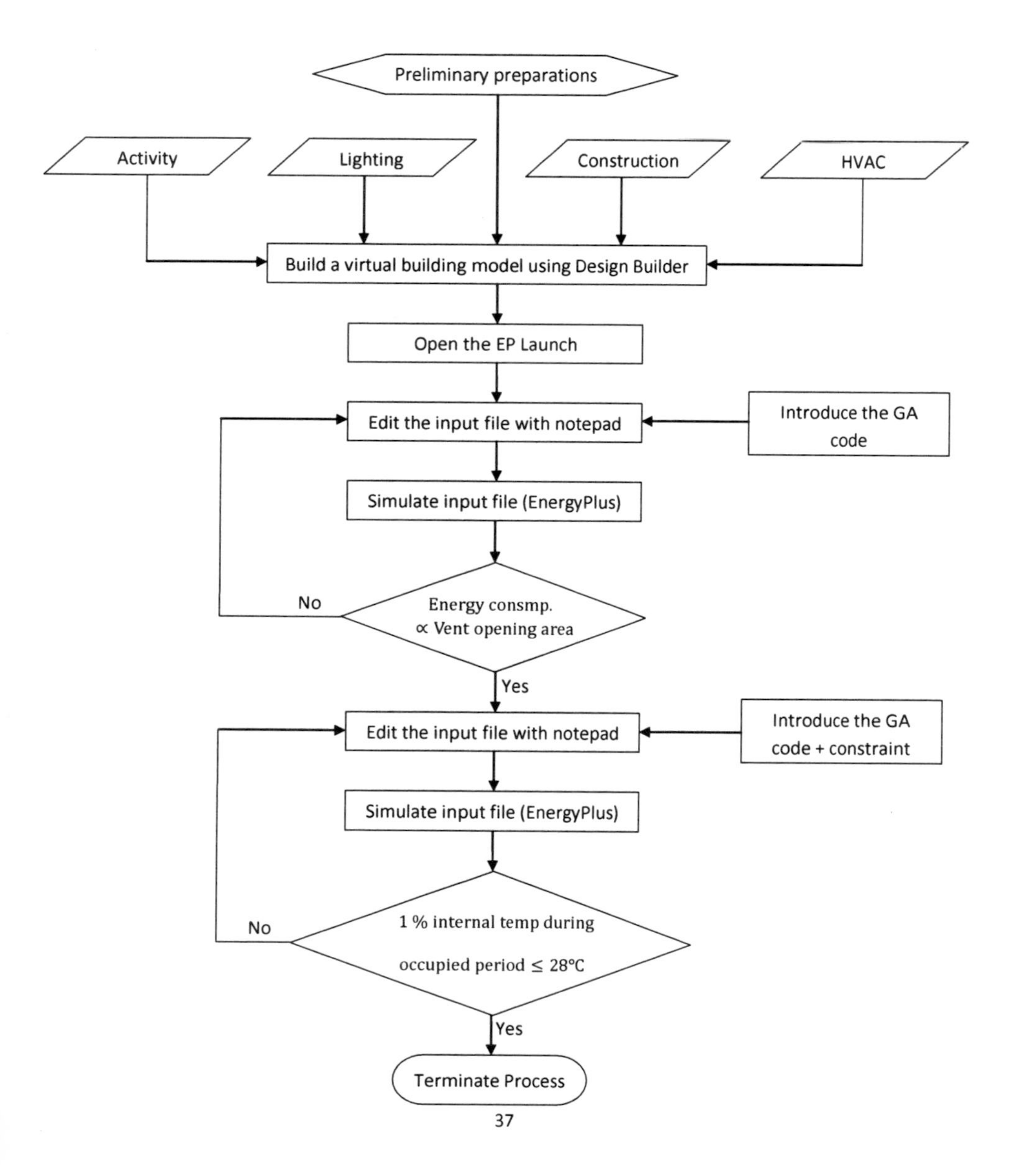

3.5.1. Preliminary Preparations

The preliminary preparations include the installation of DesignBuilder version 2.0.4.002 which can be downloaded with a 30days trial. For simulations, EnergyPlus version 3-1-0 was installed which can be downloaded for free from the internet as well. It needs to be noted that it is crucial to install the stated versions as different versions of DesignBuilder and EnergyPlus are not fully compatible with each other and may hinder editing of the source code.

3.5.2. Building the virtual model

To start optimizing vent openings we need a building on which to build the vents on. The task was to construct the virtual model of the building in the provided 3D model space. Firstly, the building's exterior walls were constructed which resulted in the formation of blocks, which were then further sub-divided into zones as internal walls were added. Once zones were created, then any particular wall could be selected on which windows and vents of any dimension could be constructed. At this point the window and vents were drawn as specified by the competition briefing document but the vent opening areas were 3% of the floor area i.e. the maximum recommended size.

3.5.3. Defining the properties of the Building

Once the building model was constructed, the activity, construction, openings, lighting and HVAC tabs in the DesignBuilder software were used to define the parameters of the building.

Activity Tab						
Floor	Density (ppl/m2)	Gain from office equipment (W/m2)	Heating Setpoint Temperatures (°C)	Heating set back (°C)	Cooling Setpoint Temperatures (°C)	Cooling set back (°C)
Ground	0.18	15	22	12	24	28
First	0.09	15	22	12	24	28
Second	0.27	15	22	12	24	28

Table 3. 1: Activity Inputs for the building

Construction Tab		
Type	Outside to inside	Thickness (mm)
External Wall	Brickwork (outer leaf)	102
	MW Stone Wool (rolls)	200
	Concrete Block (medium)	140
Inter Wall	Concrete Block (medium)	140
Roof	Asphalt	10
	UF Foam	160
	Concrete Block (medium)	140
Ground Floor	London Clay	100
	Opaque hole	100
	Cast Concrete	100
	UF Foam	100
	Screed	75
Internal Floor	Screed	75
	MW Stone Wool (rolls)	25
	Cast Concrete	150

Table 3. 2: Construction Inputs for the Building

Openings Tab		
Type	Outside to inside	Thickness (mm)
External Glazing	Clear Glass	6
	Argon Gas	16
	Clear Glass	6
Internal Glazing	Toughened	10
	Air Cavity	50
	Laminated	6
Vents	Grille, small, light slats	-

Table 3. 3: Openings Inputs for the Building

The lighting tab was used to define the lighting energy of $25W/m^2$. The HVAC tab was used to define the services installed in the building as shown below.

HVAC Tab	
Mechanical Ventilation	Off
Heating	On
Cooling	Off
DHW	On
Natural Ventilation	On

Table 3. 4: Services Inputs for the Building

The definition of these parameters ended the task of modeling the building in DesignBuilder. A snapshot of the completed building model is shown in the following figure.

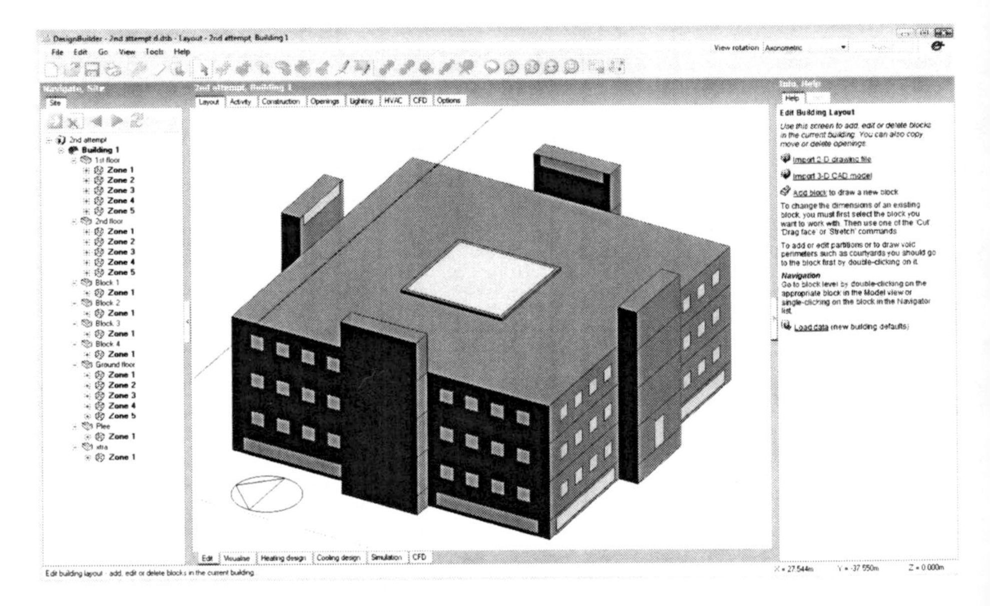

Figure 3. 3: A snapshot of the building model in DesignBuilder.

3.5.4. Running simulations within DesignBuilder

Once the model of the building was constructed, it was time to run simulations on it as it would help us in knowing if there were any errors in the model so that they could be rectified. The results obtained from the simulation would also give us an idea of the thermal performance of the building. Looking at the thermal performance it could be decided how high internal temperatures reach without stack-assisted natural ventilation and if there was any need for mechanical ventilation. For the purpose of running simulations on the building model, DesignBuilder has an in-built tab called 'Simulation' for exporting the model to EnergyPlus with the associated weather data file. After EnergyPlus is done with running simulations on the model, the results are sent back to the user without compelling them to ever leave the DesignBuilder window.

3.5.5. Editing the IDF file

In order to introduce the GA optimization code into the IDF file, the IDF file needed to be edited. This was done by the 'view the input data' button in DesignBuilder which would bring up the EP Launch tool. The EP Launch tool had a text editor button which would open the IDF file in notepad.

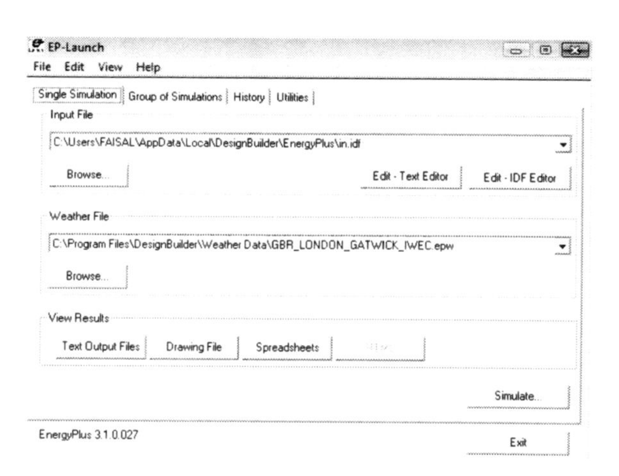

Figure 3. 4: The EP Launch interface

The IDF file consists of numerous modules such as the building materials and openings etc which help in defining the building model. Scrolling down the IDF file we find the 'FENESTRATIONSURFACE: DETAILED' module. This module consists of information about the windows, vents and holes used in the construction of the model. Each one of these components is followed by its properties which can be edited and then saved as a new input file.

We were interested in varying the opening areas of the vents so that we could find out which opening area would be the optimum when it came to meeting the requirements of the building. This could be done by varying the positions of the four vertices which defined the four corners of the vents in the 3D model space. But doing this manually was not possible as manually changing the four vertices, each of which were defined by 3 co-ordinates (in the 3D model space) alone, was just not feasible. An additional problem was the task of simulating each change we made in the IDF file and then comparing the results with our criteria. All this required for a computer code which could not only change the opening areas of the vents but also find the optimum result for us. This is where we brought in the optimization code known as the Genetic Algorithm.

3.5.6. The Genetic Algorithm Optimization

Professor J.A.Wright of Loughborough University UK wrote up a code which could make GA vary the positions of the vertices of vents in the idf file. The co-ordinates which defined the four vertices of the vents in the original IDF file were replaced with the GA code. The transformation is shown in the following figure.

```
! Vent, 4.177m2
FenestrationSurface:Detailed,
  H_1945_5_0_0_0_0_Vent,  !- Name
  Door,                   !- Surface Type
  5,                      !- Construction Name
  P_1945_5_0_0,           !- Building Surface Name
  H_1952_4_0_10000_10002, !- Outside Boundary Condition Object
  0,                      !- View Factor to Ground
  ,                       !- Shading Control Name
  ,                       !- Frame and Divider Name
  1,                      !- Multiplier
  4,                      !- Number of Vertices
  17.72136327,            !- Vertex 1 X-coordinate {m}
  -6.9614495346,          !- Vertex 1 Y-coordinate {m}
  3.162,                  !- Vertex 1 Z-coordinate {m}
  22.71736327,            !- Vertex 2 X-coordinate {m}
  -6.9614495346,          !- Vertex 2 Y-coordinate {m}
  3.162,                  !- Vertex 2 Z-coordinate {m}
  22.71736327,            !- Vertex 3 X-coordinate {m}
  -6.9614495346,          !- Vertex 3 Y-coordinate {m}
  3.998,                  !- Vertex 3 Z-coordinate {m}
  17.72136327,            !- Vertex 4 X-coordinate {m}
  -6.9614495346,          !- Vertex 4 Y-coordinate {m}
  3.998;                  !- Vertex 4 Z-coordinate {m}
```

```
! Vent, 4.177m2
FenestrationSurface:Detailed,
  H_1945_5_0_0_0_0_Vent,  !- Name
  Door,                   !- Surface Type
  5,                      !- Construction Name
  P_1945_5_0_0,           !- Building Surface Name
  H_1952_4_0_10000_10002, !- Outside Boundary Condition Object
  0,                      !- View Factor to Ground
  ,                       !- Shading Control Name
  ,                       !- Frame and Divider Name
  1,                      !- Multiplier
  4,                      !- Number of Vertices
BOP_WidthX1_20.22_5,      !- Vertex 1 X-coordinate {m}
  -6.9614495346,          !- Vertex 1 Y-coordinate {m}
BOP_HeightZ1_3.6,         !- Vertex 1 Z-coordinate {m}
BOP_WidthX2_20.22_5,      !- Vertex 2 X-coordinate {m}
  -6.9614495346,          !- Vertex 2 Y-coordinate {m}
BOP_HeightZ1_3.6,         !- Vertex 2 Z-coordinate {m}
BOP_WidthX2_20.22_5,      !- Vertex 3 X-coordinate {m}
  -6.9614495346,          !- Vertex 3 Y-coordinate {m}
BOP_HeightZ2_3.6,         !- Vertex 3 Z-coordinate {m}
BOP_WidthX1_20.22_5,      !- Vertex 4 X-coordinate {m}
  -6.9614495346,          !- Vertex 4 Y-coordinate {m}
BOP_HeightZ2_3.6;         !- Vertex 4 Z-coordinate {m}
```

Figure 3. 5: Before (left) and after (right) of the edited IDF file

The red colored font indicates where the original values have been replaced by the GA code. It may be noticed that some co-ordinates have not been edited as the vent corners as supposed to change their position only in two dimensions. Once the editing of the idf file was done it was saved as a new idf file in the working directory of the GA program. Then an exe program in MS-Dos was run which scans the new idf file for all the statements starting with the characters 'BOP' (Building optimization program). All the BOP statements would be replaced automatically by integers defining new positions for the vertices of the vent. The range in which the co-ordinates were to be varied was defined by the integers that followed the BOP characters. These integers were assigned ranges in the optimization input file which is also located in the GA program working directory. The optimization input file also defined the number of evaluations (simulations) we wanted before the GA was terminated and also the population size. Once the GA was terminated two spread sheets files were generated in the working directory of the GA program. One spreadsheet file contained all the solutions generated while the second spreadsheet contained only the optimum solution.

3.5.7. Test 1

Now that we were able to vary the opening areas of the vents in the idf file we needed to see if our model also responded accordingly. For this reason we ran the GA on our edited idf file with only the annual energy consumption (heating) as the output. Our hypothesis was that as the vent opening areas should decrease, the need for heating the building would decrease as well during winter. This would in turn result in lower annual energy consumption which was our test1. If our hypothesis is right then we should see direct proportionality between the vent opening areas and the annual energy consumption in the results. As this was just a check if our hypothesis was correct and there was no interest in optimization at this point thus we only ran 100 evaluations. This check was repeated three times keeping the evaluations size constant with the value 100 but varying the population size as 5, 10 and 20. Test 1 was carried out for the following purposes:

1) To see if the building model changed its attributes as the GA made changes in the IDF file.
2) To see if our hypothesis of $energy\ consumption \propto vent\ size$ was correct or not.
3) To observe the behavior of GA optimization as the population size is varied.

3.5.8. Test 2

With the success of test 1, the next step was to optimize the vent openings but this time with a constraint. The constraint is the same as the criteria that was stated by the competition brief. The constraint was to search for an optimum size of vent openings which would result in the internal temperatures being below 28°C for no more than 1% of the occupied period round the year. Currently the occupancy pattern makes the building being occupied for 3132 hours round the year, making the allowance of the internal temperature to exceed for no more than 31.32 hours an year.

3.5.9. The Results

As mentioned earlier the after the optimization is terminated, we would obtain two spreadsheet files in the working directory of the GA program. These would contain configurations of varying sizes of vent openings together with any criteria or in GA terms 'fitness' we define in the idf file. Let us now take a look at some of our findings in Chapter 4.

Chapter Four - Results and Analysis

4.1. Introduction

As explained in chapter 3, with the help of EnergyPlus simulations and Genetic Algorithm, an optimum cross-sectional area of the vent openings is investigated in this research effort. This chapter presents the results obtained after running the simulations in accordance with the methodology presented in the previous chapter. The aim is to avoid summer time over heating and reduce annual energy consumption by optimizing the vent openings in a naturally ventilated building. According to the criteria of the competition and CIBSE guide A, for a naturally ventilated office in the UK, the internal temperatures should not exceed for more than 1% of the occupancy period. For the selected office building, working hours are from 7am to 6pm thus making a total of 3133 hours annually. This makes an allowance of only 31 hours round the year for the internal temperatures to go above 28°C.

4.2. Investigation 1

This investigation is based on figuring out if changes in the cross-sectional area of the vent openings have any effect on the temperatures and annual energy consumption of the building.

4.2.1. Is there a need for natural ventilation?

The first and most basic question to answer before any further research is carried out is to see if the building could do without stack-assisted natural ventilation. For this reason the building form and its components were kept the same, yet the inlet and outlet vent openings on each level of the building were completely removed in order to see the response of the building. This was done by simply deleting all the vent elements from the building model in DesignBuilder. Yet it is not practically possible to have no ventilation in a building thus natural ventilation due to windows was added to the model. The windows were kept open 5% of their opening area throughout the occupancy period i.e. 7am to 6pm. The thermal behavior of the building is shown in the following graph.

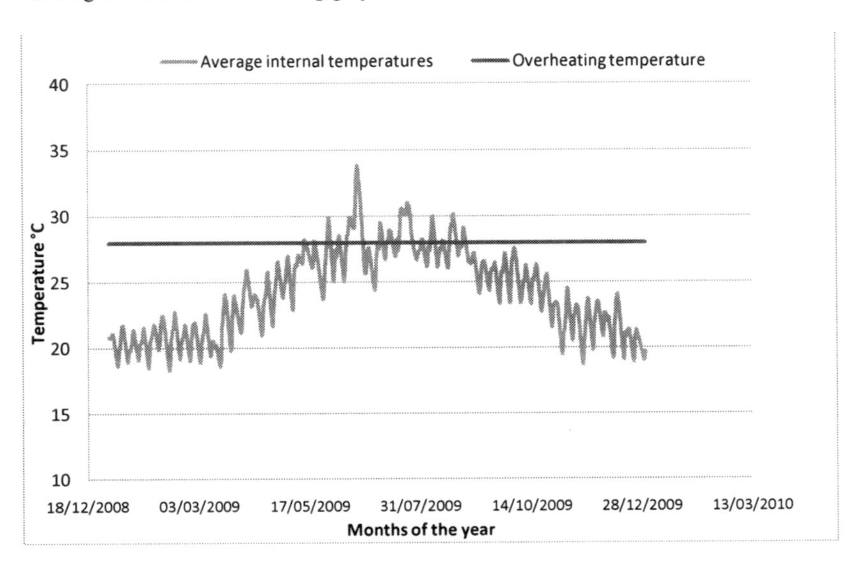

Figure 4. 1: Dry resultant temperature without stack-assisted Natural Ventilation.

It can be seen from figure 4.1 that the dry resultant temperature is above 28°C for quite some time round the year. Tabular data show that the dry resultant temperatures are higher than 28°C for more than 87 hours of occupied period. It needs to be reminded at this point that according to CIBSE Guide A, the dry resultant temperatures for our building should not exceed 28°C for more than 31 hours during the occupied period. Thus there is a need of stack-assisted natural ventilation in order to bring the internal temperatures down and hence the vents cannot be completely removed in hopes of reducing energy loss.

4.2.2. Is there a need for mechanical ventilation?

Once it was clear that stack-assisted natural ventilation was needed for reducing the internal temperatures, the next question was if natural ventilation alone could meet the summertime overheating criteria. For this reason the building model in DesignBuilder was installed with vent openings having the maximum recommended size. This meant that the cross-sectional area of the vent opening was made 3% of the floor area making each vent opening having a cross-sectional area of approximately $4m^2$. The dry resultant temperatures for such a case are shown in the following graph.

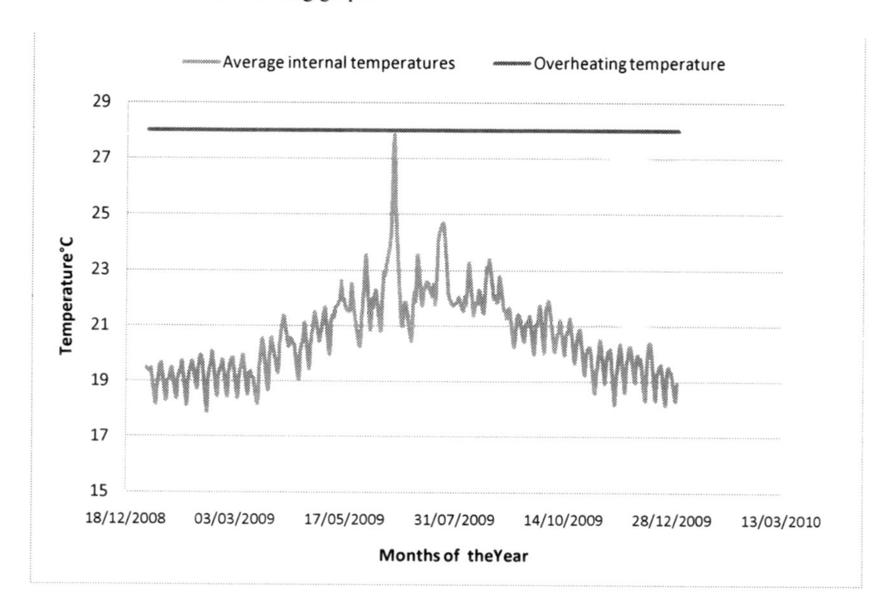

Figure 4. 2: Dry resultant temperatures with vent opening area being 3% of the floor area

From figure 3.2 we observe that the maximum recommended size of vent opening allows the dry resultant temperatures to cross the 28°C temperature limit for only 1 hour throughout the year even though it has the allowance to cross it up to 31 hours. This shows that the assistance of mechanical ventilation will not be necessary in our test building and that the natural ventilation is enough to meet the overheating criteria.

4.2.3. Is Annual Energy Consumption affected by variation in vent opening areas?

To answer this question we take a look at two scenarios. The first scenario being that the vent openings are sized 1% of the floor area while the second scenario is that vent openings are sized 3% of the floor space. For both the scenarios the monthly ventilation rates and zone heating levels are recorded. 'Zone heating' is the energy supplied by local room heaters and reheat coils to maintain room internal heating setpoint temperature.

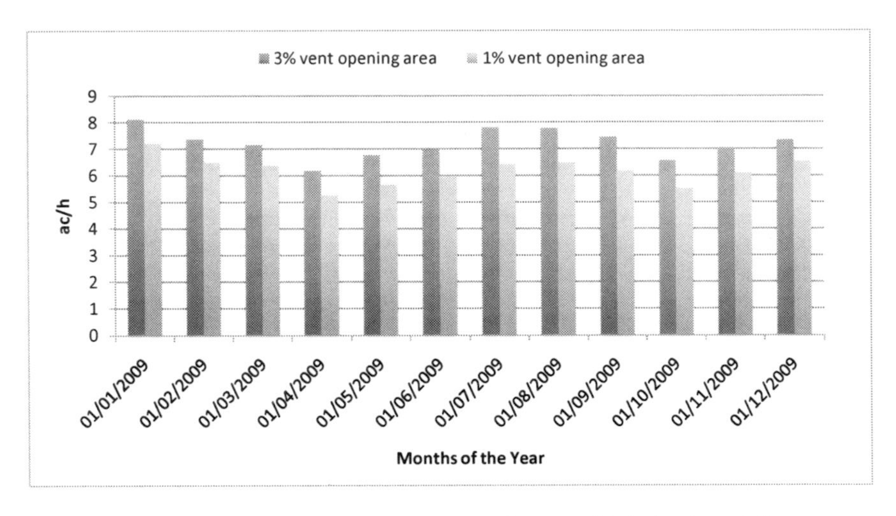

Figure 4. 3: Affect of Vent opening variation on ventilation rates (Tabular data: Appendix B)

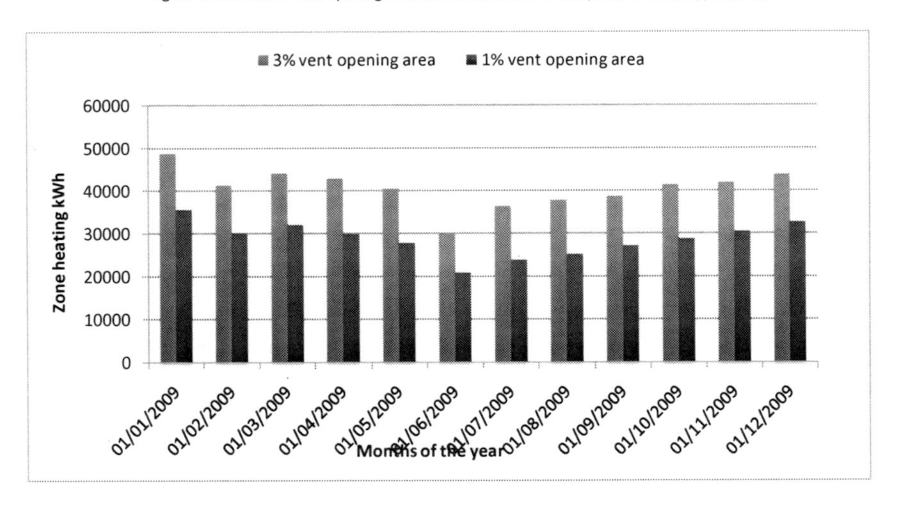

Figure 4. 4: Affect of Vent opening variation on zone heating (Tabular data: Appendix B)

Figure 3.3 suggests that as the vent opening areas are increased the amount of ventilation also increases, causing the heaters to supply more energy to maintain the internal set point temperatures known as zone heating. Figure 3.4 shows the variation in zone heating as the vent opening area is decreased from 3% to 1% of the floor area. Thus the overall result of the above investigation is that energy consumption is affected by variation in the cross-section area of vent openings.

4.3. Investigation 2

Investigation 2 is that stage of the project where we introduce Genetic Algorithm into the input data file, created from the building model. This investigation will help in answering the following questions.

- With every new evaluation, is GA capable of changing the model of the building?
- Is our hypothesis of energy consumption being directly proportional to the area of the vent opening correct?
- Does our GA algorithm converge towards the optimum result?
- Is energy consumption more sensitive to the inlet or the outlet vent opening area?

Answering the first question is easy. With every new evaluation, GA generates a new IDF file in its 'Archive' directory. Upon opening each of these idf files one can see the co-ordinates of the vent vertices changing. This confirms to us that our optimization algorithm is in fact changing the attributes of the vent openings in the building model.

To answer the proportionality question, GA optimization was run varying vent opening with the annual energy consumption of the building as the fitness criteria. If our hypothesis is correct then for each new generation, GA should try to try to converge towards the smallest value of vent opening area in order to reduce the energy consumption. A graph was plotted between an equivalent vent opening area (which was the average of both the inlet and outlet vent opening areas) and the annual energy consumption. This graph is shown in the following figure.

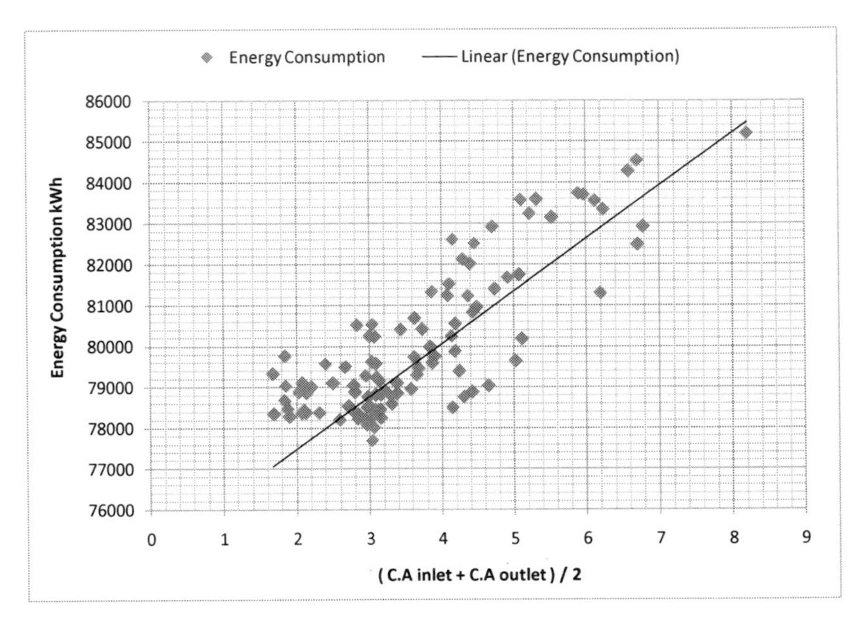

Figure 4. 5: Equivalent Vent Opening Areas versus Annual Energy Consumption

The blue dots in figure 3.5 represent each solution that GA generated. We see that the dots are quite randomly placed yet the overall behavior of these blue dots is represented by the black line. The black line suggests a linear and a directly proportional relation between the equivalent vent opening area and energy consumption. We also observe the concentration of the blue dots increasing near the lower values of opening areas. This suggests that GA is in fact converging towards smaller vent opening areas with the aim of reduction of energy consumption. But the question still remains why such high randomness in the solutions generated? And according to our hypothesis GA should have reduced the vent openings to the smallest value possible but we see it still has not converged to the smallest possible opening areas. Why is it so?

There could be two possible answers to these questions. The first answer could be that GA was varying the inlet and outlet vents with different openings areas, during each generation. This we know is true as we had assigned two different and independent variables for changing the areas of the inlet and outlet vents in the GA input file. This results in random configurations of inlet and outlet vents being generated independent of each other and thus causes the high randomness seen in figure 3.5. The second answer could be that energy loss is dependent more on, either the inlet or outlet vent opening area. This is worth investigating and so we try to relate the cross sectional area of the inlet and outlet vent separately with the annual energy consumption.

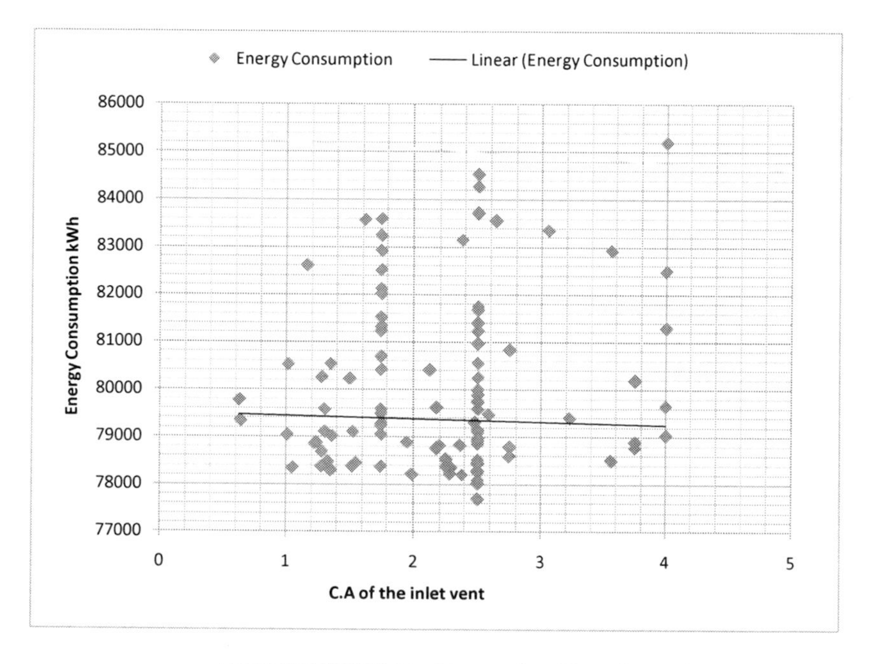

Figure 4. 6: Inlet Vent Opening Area versus Annual Energy Consumption

The graph above makes it clear that variation in the cross-section of the inlet vent opening does not have a significant effect on the annual energy consumption. We can see that the GA also detects this non-dependency of the energy consumption on the cross-sectional area of the inlet vent opening. That is why GA generates, during most of its evaluations, intermediate values for area of the inlet vent.

On the other hand if we look at the relation between the cross sectional area of the outlet vents and energy consumption, we obtain the following graph.

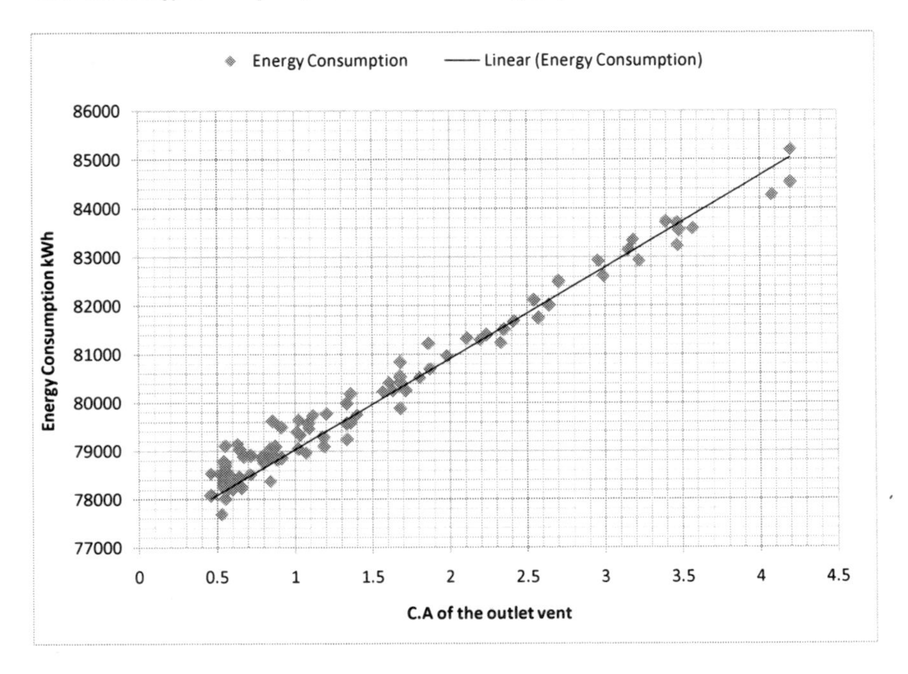

Figure 4. 7: Outlet Vent Opening Area versus Annual Energy Consumption

It is observed from figure 3.6 that not only is energy consumption very sensitive but is also directly proportional to changes in the outlet vent opening area. We also see that GA has converged towards the smallest possible opening areas of the outlet vent. This is exactly what we wanted our optimization algorithm to do at this stage.

4.4. Investigation 3

Figure 3.6 raised another question as to why the energy consumption was more sensitive to the outlet vent opening than the inlet vent opening. Till now the change we had been observing in the energy consumption was because of a component known as the zone heating as described earlier. Zone heating is dependent on the amount of ventilation occurring in the building. The higher the ventilation rate the higher will be the zone heating and thus annual energy consumption. So could it be that the outlet vent was the major factor in regulating the ventilation rates in the building thus causing the major difference in the zone heating? To investigate the answer to this question we studied the ventilation rates for the following four cases.

Case Number	Opening Area as a percentage of the floor area served	
	Inlet Vent	Outlet Vent
Case 1	1%	1%
Case 2	3%	1%
Case 3	1%	3%
Case 4	3%	3%

Table 4. 1: Four cases of investigation

The followings graph shows the ventilation rates throughout the year for each of the cases mentioned.

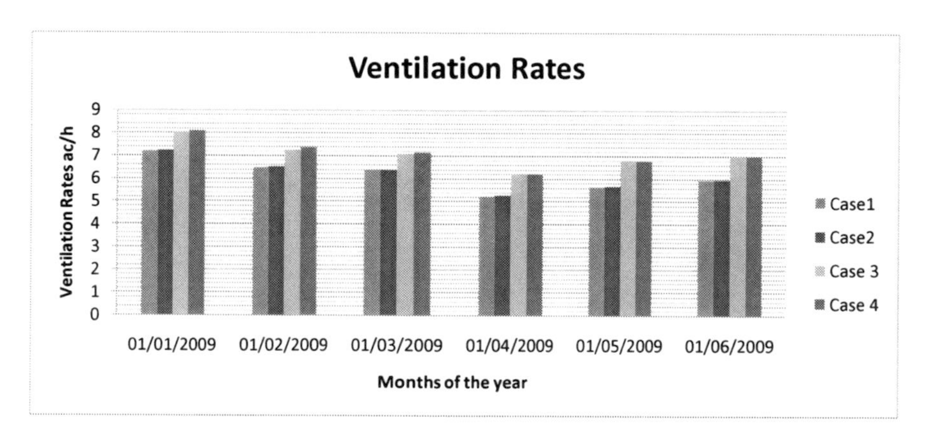

Figure 4. 8: Ventilation rates for each of the four cases (Tabular data: Appendix C)

Case 3 readings in figure 3.7 suggests that reducing the inlet vent to 1% does not show a major decrease in the ventilation rates if the outlet vent is kept at 3%. So much so that case 3 results are almost similar to case 4 results. On the other hand, reducing the outlet vent opening area to 1% causes a dramatic decrease in ventilation rates, almost as same as case 1. This confirms that the outlet vent is what regulates the ventilation rate. This suggests that ventilation in the building is working more on a 'pull' rather than a 'push' system. A pull system means that higher flow rates out of the building cause higher flow rates into the building. Lower flow rates out of the building due to smaller outlet vents will cause lower flow rates into the building no matter how much the size of the inlet is increased in the 1 to 3% range.

This finding is important, as now the inlet vents can be kept the smallest recommended size in order to avoid noise exchange amongst the different levels of the building via the lightwell, which is a common problem in naturally ventilated buildings having a lightwell.

4.5. Investigation 4

It is important to understand the nature of the Genetic Algorithm before we use it for optimization purposes. For this reason the GA optimization is run three times by keeping the evaluations size 100 but varying the population size as 5, 10 and 20. This is done by varying the population size value in the GA input file. Let us now take a look at the results obtained from the three simulations.

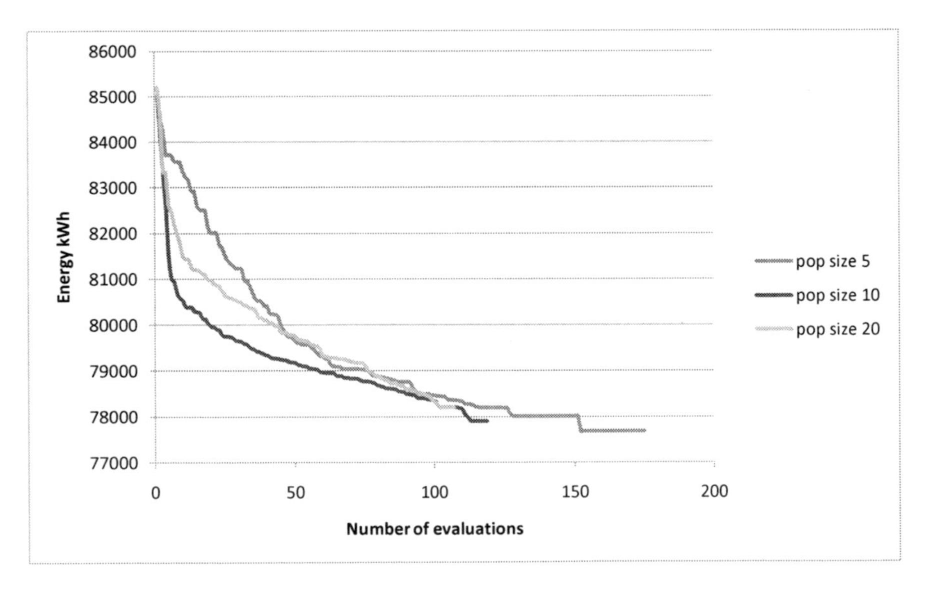

Figure 4. 9: GA convergence with variation in Population size

The preceding graph suggests that a population size of 5 presents us with the least energy consumed but takes considerably longer for it to reach that optimum solution. A population size of 10 gives us a pretty optimum result as well converges towards the optimum result more rapidly. Population size of 20 though, not only takes longer time to converge but also results in the highest energy consumption. Due to computational and time limitations we want our optimization algorithm to converge quickly towards the optimum result thus for finding the optimum vent size between 1% to 3%, we will be using a population size of 10 while the evaluation size will be increased to 400 for a more thorough search for an optimum result.

4.6. Investigation 5

Investigation 5 is the main objective of this whole research. It is focused on practically implementing the GA tool on the test building of ours and finding an optimum size for its vents in order to meet the criteria of the brief whilst still keeping a low carbon footprint. This time the GA code was introduced into the IDF file but with a constraint of keeping temperatures below 28°C for as much time as possible. The results were presented in spreadsheet files. In the spreadsheet files the first four columns represented different configurations of widths and heights for the inlet and outlet vents respectively. The following columns expressed, for each level of the building, number of hours the dry resultant temperatures were above 28°C for each configuration. These values were presented in such a way that to get the actual number of hours the internal temperatures were above 28°C one had to add 31 to it, 31 being our allowable number of hours above 28°C. For example, if the column contained a value of -18 then add 31 to it i.e. -18+31 = 13 hours above 28°C in that year. Also the energy consumption for each configuration of inlet and outlet vent openings was reported.

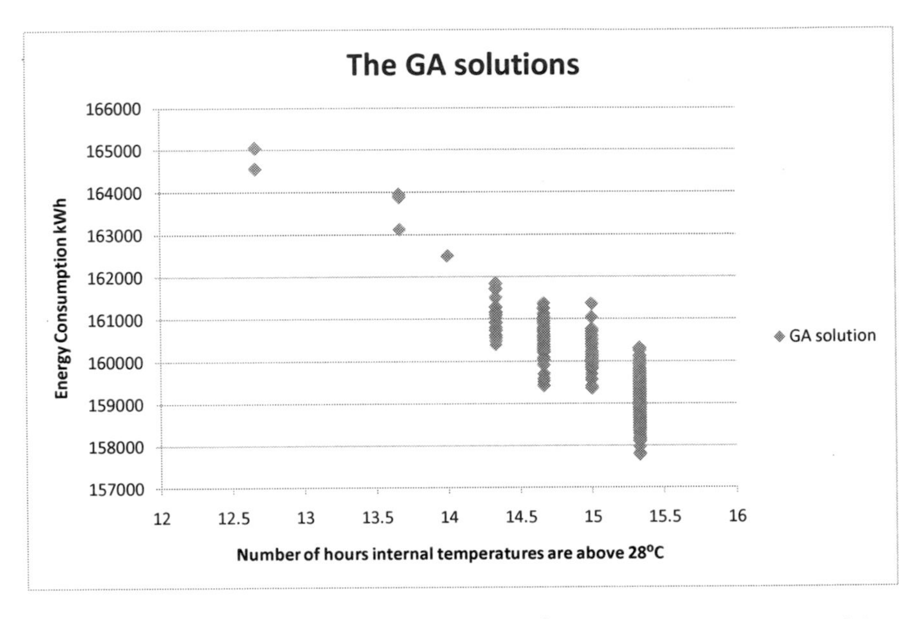

Figure 4. 10: Number of hours the dry resultant temperature was above 28°C versus energy consumption for the 400 solutions GA generated

The graph above shows the solutions GA generated. It is found from these solutions that for any number of hours for which internal temperatures are above 28°C, there can be different configurations of vent openings. This means that different sizes of inlet and outlet vents can result in the same cooling strategy. The best option would be to select the smallest vent opening configuration for the desired cooling strategy and thus have smaller energy consumption annually.

The optimum result obtained from running the 400 evaluations was:

Inlet Vent		Outlet Vent		Hours Above 28°C			Energy
width	height	width	height	Level 1	Level 2	Level 3	kWh
1.55	0.25	2.55	0.23	10	19	17	157774.7

Table 4. 2: The optimum configuration

The table above shows that for the stated dimensions of inlet and outlet vents the 28°C temperature limit is only crossed for 19 hours in level 2, whereas we have the allowance to cross it for up to 31 hours. Plus the energy consumed is 157774.7 kWh for this configuration while the maximum recommended sizes of vents would have caused an energy consumption of 166411.9 kWh. Thus the GA optimization tool has not only met the criteria of summer time overheating but has also resulted in saving 8637.2 kWh of energy per year.

Chapter Five – Conclusions and Recommendations

Chapter Five - Conclusions and Recommendations

5.1.Introduction

This chapter presents a summary of the findings, followed by conclusions, limitations of the study and recommendations for future research.

The simplest naturally ventilated building can fail in avoiding summer time overheating and high energy consumptions if it is not properly designed. Good ventilation design needs significant consideration in order to achieve the overheating criteria of CIBSE Guide A for naturally ventilated buildings. Correct vent opening design is needed to exhaust the correct amount of stale air to the outside environment to maintain thermal comfort levels during summers yet avoiding high heat losses during winters. Thus, vent opening design is one important aspect of optimum control of natural ventilation. In this research, we managed to the goals mentioned in Section 1.3. The contributions of this research are discussed in the following sections.

5.2. Contributions of this research

Linking Genetic Algorithm with virtual building models

Genetic Algorithm is one of the best and commonly used optimization tools today. However, using it to optimize a particular feature of a virtual building model needed a link to be formed between the building model and GA. This was achieved by using DesignBuilder as the modeling software and then creating an IDF file from that building model. The original attributes of the feature which needed optimization were replaced by the GA code. GA was then able to randomly change the attributes of the feature and then converge towards a particular configuration based on a particular fitness.

The nature of GA was better understood

Using GA as the optimization tool gave us an insight into how GA worked. GA used an input file to run optimizations in a particular manner. The input file consisted of two important variables known as the 'evaluation size' and the 'population size'. It was understood that higher the value of evaluation size the more thorough the solution space will be searched for the optimum result. As far as population size was concerned, it was found that increasing the population size results in a more optimum result. But the increase in population size also meant GA would take a longer time to reach that optimum solution.

Effect of variation in vent openings on internal temperatures

For this investigation the virtual model was manually constructed with two different configurations of vent openings. Having vent openings the size of 1% caused the internal temperatures to be 87 hours above 28°C. Changing the vent openings to 3% of the floor area, resulted the internal temperatures to exceed the 28° temperature limit for only one hour.

Validation of the vent opening and energy consumption proportionality

After GA was setup, it was used to optimize the vents of the virtual building model and also report on the annual consumption of energy. It was found that energy consumption increased with increase in vent opening area and vice versa. Plotting a graph between the vent opening areas and annual energy consumption showed direct proportionality between the two variables.

Energy consumption sensitivity analysis

Carrying out a sensitivity analysis of the energy consumption with the inlet and outlet vent separately resulted in an interesting finding. It was observed that energy consumption did not vary with the variation of the inlet vent. On the other hand the outlet vent varied the energy consumption significantly. Higher ventilation rates causes higher energy loss during winter and hence causes higher energy consumption due to zone heating. After the analysis of ventilation rates of the building, it was found that it was the outlet vent opening area that controlled the flow through the building. The higher the outflow the higher was the inflow. And so the inlet vents opening areas should be close to 1% of the floor area served in order to avoid noise exchange between different levels of the building. As for the outlet vent opening areas, they should be reduced until they start to fail the summer time overheating criteria.

Achievement of a new design optimization tool for buildings

With the achievement of linking GA with the virtual model of the building a new design tool was presented to assist designers in making correct decisions. This design tool would help designers optimize the vent openings for any cooling criteria that they choose to implement in their naturally ventilated buildings. They would also be given more than one configuration, if existing, of vent openings in order to achieve the same cooling strategy if there were limitations to the size of vent openings in the building. The designers would also be given a value of the annual energy consumption as a consequence of choosing a certain configuration. Using it on our building and optimizing the vent openings, saved 8637.2 kWh of energy consumption due to zone heating.

5.3.Limitations

The most influential limitation of this research is that GA evaluations were limited due to computational and time consuming reasons. Having higher evaluation size could have resulted in a different and more optimum configuration of vent openings.

The study focused on one criteria of CIBSE Guide A about avoiding summer time overheating and reducing energy consumption. Yet there are other criteria or constraints which can be incorporated into the GA such as carbon dioxide levels so that the configuration it recommends is even better.

A third limitation of this research is that it looks only at the sizing of the inlet and outlet vents. While this limitation is addressed in the recommendations for future research, the findings do not support any conclusion on how the vents should be modulated.

5.4.Future Recommendations

This research has looked in detail at the optimum sizing of vents but it is of vital importance to plan an optimum strategy for modulating the vents as well. A particular tool of EnergyPlus allows the control of natural ventilation through exterior and interior openings. If the natural ventilation control is based on temperature then all the vents are opened if $T_{zone} > T_{out}$ and $T_{zone} > T_{set}$. For the modulation of the vents, the user is intended to input the following values:

- 'Minimum vent opening factor' the value of which may be from zero to 1.0.
- 'Indoor and outdoor temperature difference lower limit for maximum venting open factor'. This value may be from 0°C to 100°C but should be less than the value specified for the following field.
- 'Indoor and outdoor temperature difference upper limit for minimum venting open factor'. The value must be greater than 0°C and must be greater than the value specified in the previous field.

The following figure explains what each of these inputs represent.

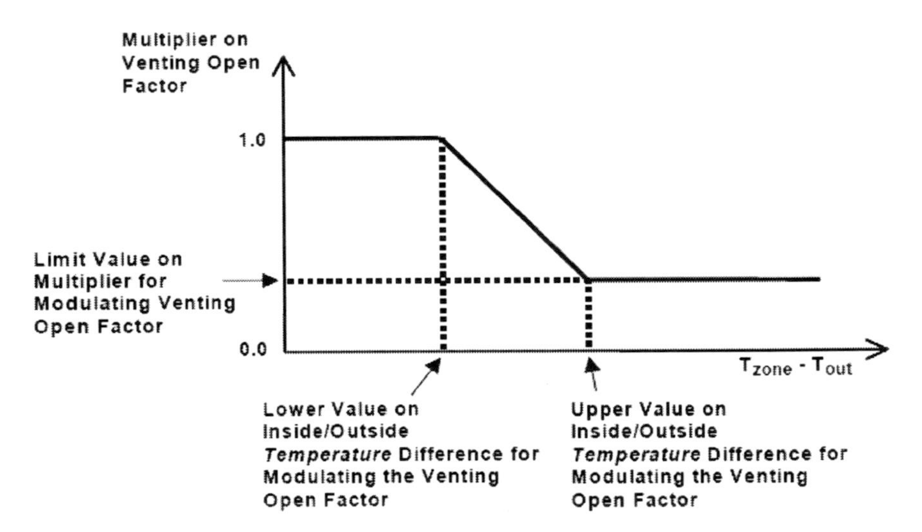

Figure 5. 1: Modulation of vents according to inside-outside temperature difference. Source: EnergyPlus Documentation.

If GA is used to vary the above inputs, an optimum control strategy can be found for modulating the vents. Further investigation can be executed in order to find why those inputs, which GA recommends, are optimum?

APPENDIX A

The Building Model

Appendix A

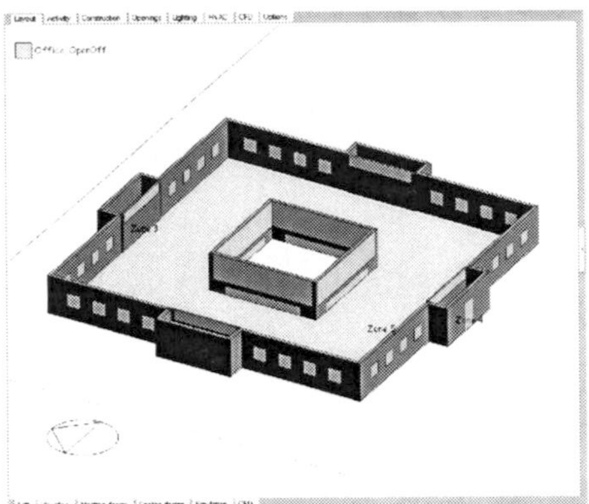

(Left) Inside view

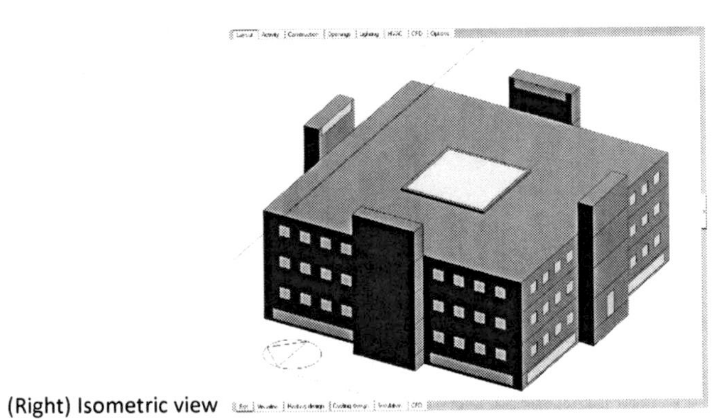

(Right) Isometric view

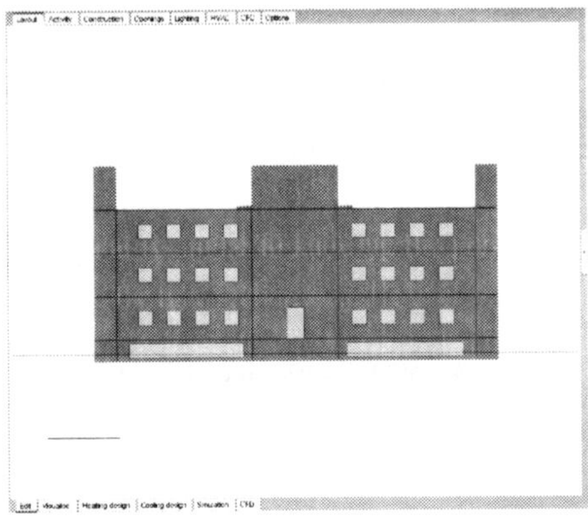

(Left) Front view

APPENDIX B

Ventilation rates and Zone heating for 3% and 1% vent opening

Date/Time	3% vent opening ac/h	1% vent opening ac/h
1/1/2009	8.103358	7.194801
2/1/2009	7.365813	6.486305
3/1/2009	7.145821	6.372919
4/1/2009	6.178621	5.232244
5/1/2009	6.774014	5.644677
6/1/2009	6.964326	5.934045
7/1/2009	7.801935	6.407325
8/1/2009	7.769489	6.47244
9/1/2009	7.444398	6.166026
10/1/2009	6.554823	5.514553
11/1/2009	7.036637	6.104502
12/1/2009	7.331817	6.538909

Date/Time	3% vent opening Zone/Sys Sensible Heating kWh	1% vent opening Zone/Sys Sensible Heating kWh
1/1/2009	48760.99	35641.02
2/1/2009	41298.1	30073.19
3/1/2009	44176.8	32085.17
4/1/2009	42942.81	29863.7
5/1/2009	40525.72	27820.91
6/1/2009	30088.76	20870.27
7/1/2009	36491.76	23893.7
8/1/2009	37890.7	25218.06
9/1/2009	38728.27	27152.16
10/1/2009	41477.15	28809
11/1/2009	41938.41	30507.12
12/1/2009	43786.82	32676.09

APPENDIX C

Ventilation Rates for the Four Cases

Appendix C

Date/Time	Case 1 Nat Vent	Case 2 Nat Vent	Case 3 Nat Vent	Case 4 Nat Vent
	ac/h	ac/h	ac/h	ac/h
1/1/2009	7.194801	7.219443	8.008546	8.103358
2/1/2009	6.486305	6.512187	7.245059	7.365813
3/1/2009	6.372919	6.361799	7.092354	7.145821
4/1/2009	5.232244	5.265438	6.194944	6.178621
5/1/2009	5.644677	5.656045	6.766191	6.774014
6/1/2009	5.934045	5.960982	7.011079	6.964326
7/1/2009	6.407325	6.427383	7.828084	7.801935
8/1/2009	6.47244	6.424338	7.70227	7.769489
9/1/2009	6.166026	6.172358	7.324924	7.444398
10/1/2009	5.514553	5.496467	6.596817	6.554823
11/1/2009	6.104502	6.10836	6.99299	7.036637
12/1/2009	6.538909	6.559815	7.235068	7.331817

References

- BRITISH STANDARDS INSTITUTION (1995) "Moderate thermal environments - determination of the PMV and PPD indices and specification of the conditions for thermal comfort", EN IS0 7730:1995. ISBN580223470
- BRE (1999) "Natural ventilation for offices", *Building research establishment/NatVen.*
- BRE (1998) "Night ventilation for cooling office buildings", *Information Paper IP4/98 Building Research Establishment Ltd.*
- BS5925 (1991) "Ventilation principles and designing for natural ventilation", *British Standards Institution*, London.
- CIBSE Guide B2 (2001) "Ventilation and air conditioning", ISBN 1903287162.
- CIBSE, AM10 (2005) "Natural ventilation in non-domestic buildings", *Applications Manual,* ISBN 1903287561.
- CIBSE (2000) "CIBSE guide volume H: building control systems".
- Da Graca, G. C., Linden, P. F. & Haves, P. (2004) "Design and testing of a control strategy for a large, naturally ventilated office building", *Building Services Engineering Research and Technology*, Vol. 25, 233. DOI: 10.1191/0143624404bt107oa.
- Eftekhari, M.M., Marjanovic L.D. & Pinnock D.J. (2003) "Air flow distribution in and around a single-sided naturally ventilated room", *Building and Environment,* Vol. 23, 574-581.
- EnergyPlus (2008) "EnergyPlus Engineering Reference", *EnergyPlus documentation.*
- Evans, R., Haste, N., Jones, A. & Haryott, R. (1998) "The long term costs of owning and using buildings", *Royal Academy of Engineering*, London.
- Evola, G. & Popov, V. (2006) "Computational analysis of wind driven natural ventilation in buildings", *Energy and Buildings,* Vol. 38, 491-501.
- Fanger P.O. (1972) "Thermal comfort", *McGraw-Hill.*
- Fracastoro, G. V., Mutani, G. & Perino, M. (2002) "Experimental and theoretical analysis of natural ventilation by windows opening", *Energy and Buildings,* Vol. 34, 817-827.
- Heiselberg, P. (2004) "Natural Ventilation Design", *International Journal of Ventilation*, Vol. 2, 295-312.

- Hughes BR & Ghani S.A.A Abdul (2008) "A numerical investigation into the effect of windvent dampers on operating conditions", *Building and Environment,* Vol. 44, 237-248.
- Jackman, P.J. (1999) "Air Distribution in Naturally Ventilated Offices", *BSRIA Technical Note TN 4/99,* ISBN 0860225224.
- Jiang, Y. & Chen, Q. (2003) "Buoyancy-driven single-sided natural ventilation in buildings with large openings", *International Journal of Heat and Mass Transfer,* Vol. 46, 973-988.
- Levermore, G. J. (2000) "Building energy management systems", *E&FN Spon.*
- Liem, S. H., van Paassen, A. H. C. (1997) "Hardware and controls for natural ventilation Cooling", *18th AIVC conference on ventilation and cooling*, Athens.
- Mahdavi, A. & Proglhof, C. (2008) "A model-based approach to natural ventilation", *Building and Environment,* Vol. 43, 620-7.
- Martin, A. & Fitzsimmons, J. (June 2000) "Making Natural Ventilation Work", *BSRIA Guidance Note GN 7/2000*, ISBN 0860225534.
- Martin, A. J. (1996) "Control of Natural Ventilation", *BSRIA Technical Note TN11/95* ISBN 0860224449.
- OSELAND N. (1994) "A review of thermal comfort and its relevance to future design models and guidance", *Proceedings of BEPAC Conference.* York. 205-216.
- Sandberg, M. (2004) "An alternative view on the theory of Cross-ventilation", *International Journal of Ventilation*, Vol. 2, 409-418.
- Shaviv, E., Yezioro, A. & Capeluto, I. G. (2001) "Thermal mass and night ventilation as passive cooling design strategy", *Renewable Energy,* Vol. 24, 445-452.
- Spindler, H. C. & Norford, L. K. (2008) "Naturally ventilated and mixed-mode building-Part II: Optimal control", *Building and Environment,* Vol. 44, 750-761.
- Spindler, H. C. (2004) "System identification and optimal control of mixed-mode cooling", *Thesis, (PhD). Massachusetts Institute of Technology, Dept. of Mechanical Engineering.*
- Stavrakakis, G. M., Koukou, M. K., Vrachopoulos, M. G. r & Markatos, N. C. (2008) "Natural cross-ventilation in buildings: Building-scale experiments, numerical simulation and thermal comfort evaluation", *Energy and Buildings,* DOI:1016/j.enbuil.2008.02.022.

- Van Paassen, A. H. C., Liem, S. H. & Groeninger, B. P. (1998) "Control of night cooling with natural ventilation: sensitivity analysis of control strategies and vent openings", *19th annual AIVC conference,* Oslo.
- Wright, J. A., Loosemore, H. A., Farmani, R. (2002) "Optimization of building thermal design and control by multi-criterion genetic algorithm", *Energy and Buildings,* Vol. 34, 959–72.
- Yao, R., Steemers, K. & Baken, N. (2005) "Strategic design and analysis method of natural ventilation for summer cooling", *Building Service Engineering Research and Technology,* Vol. 26, 315. DOI: 10.1191/0143624405bt139oa.